BUILDING THE IMPACT FORCE

MARINE CORPS TRANSFORMATION IN AN AGE OF CHAOS

ROBBIN LAIRD

ISBN: 979-8-9989369-5-1

Library of Congress Number: 2026904599

The cover image was created by an AI system.

Published by Second Line of Defense

Arlington, Virginia

The picture on the back cover was shot during my visit to Eglin AFB in 2012 where I am standing next to Major Bachman. Major Bachmann flew the 200th sortie of the F-35 at Eglin on August 24, 2012. I had the unique opportunity to inform him of that fact after he got out of his cockpit and then afterwards had a chance to sit down and discuss progress on the F-35 at Eglin.

I am dedicating this book to Colonel "Chip" Berke, who embodies the kind of USMC leadership that drives transformation in the face of political flak.

I first interviewed him when he was at Nellis AFB flying the F-22 as a Marine exchange pilot, a role made possible by Secretary Mike Wynne and General T. Michael "Buzz" Moseley's determination to expose other services to fifth-generation airpower.

I later asked him to join our seminars in Copenhagen and Canberra, where he shared his fifth-generation experience with audiences often skeptical of, and at times hostile to, the very change he represented.

One of my most memorable moments came at Eglin AFB, when Ed Timperlake and I visited with Secretary Wynne and witnessed a conversation between the architect of fifth-generation airpower and one of its most important operational pioneers. That exchange is reproduced in the appendix to this book.

To "Chip" goes the credit for speaking clearly and directly to the critics of fifth-generation airpower, and for reminding us how Marines lead change even when the critics would prefer that change never arrive.

"Chip" Berke presenting at the 2014 Sir Richard Williams Foundation seminar on the future of AirPower.

FOREWORD

Foreword by George Trautman, LtGen, USMC (Ret)

The United States Marine Corps has never had the luxury of preparing for the last war. From Belleau Wood to Fallujah, Marines have been forced to adapt to environments that were seldom predictable. What Robbin Laird has captured in *Building the Impact Force* is that our present moment is no different in its demands, but very different in its character.

This book describes a Corps in the midst of a deliberate, often contentious, transformation from a crisis-response force built around self-contained Marine Air-Ground Task Forces (MAGTF's) to an impact force designed from the outset to generate disproportionate effects for the joint and combined team.

Laird's vantage point is not that of a distant commentator, but of a longtime professional observer who I first encountered during the programmatic fight to deliver the MV-22 Osprey and F-35 Joint Strike Fighter to the Corps between 2007 and 2010. Across almost 20 years of work with our aviation community, from MAWTS-1 to the Marine Aircraft Wings that compose our Aviation Combat Elements, he has helped Marines turn the promise of platforms like the MV-22, F-35, UH-1Y, AH-1Z, KC-130J, and CH-53K into operational reality.

What makes this volume distinctive is its insistence on practice over theory. For example, the Steel Knight 25 exercise is not presented as a glossy validation event, but as a campaign laboratory in which I MEF and 3rd MAW wrestle with the hard problems of distributed operations: mobile, expeditionary contested logistics, fragile communication webs, emission control, authorities and risk, and the human limits of cognition in a data-saturated battlespace.

This is not a story of magic solutions. Laird is candid about the constraints that every Marine Air Task Force commander faces. Our ability to move and sustain distributed forces across long distances hinges on finite numbers of aircraft and on aviation ground support capabilities that are often stretched to the limit. Our culture of mission command must contend with the temptation to centralize when networks are rich, even as survivability and tempo demand that we push decisions and authorities down to the small, isolated nodes operating inside an adversary's weapons engagement zone.

At the same time, this book gives due weight to what remains of our decisive advantage: people and culture. In interview after interview, Laird gives voice to Marines who are not looking in their rearview mirrors at the debilitating effects of Force Design 2030 or waiting for the arrival of perfect doctrine or equipment. These Marines are experimenting in real time — refining the hub and spoke construct, rethinking how H-1s and MQ-9s can serve as low-altitude C2 and sensing nodes, leaning into the promise of AI/ML tools, building Forward Arming and Refueling Points (FARPs) hundreds of miles away from home base, and treating "digital interoperability" not as a buzzword but as a daily reality within the battle space. You also see an institution willing to adjust course through a *Campaign of Learning*: partially reversing earlier divestments in aviation and aviation ground support, preserving conventional infantry capacity alongside so called "stand-in" forces, and updating Force Design guidance in light of the operational friction being encountered in the field.

For Marines of my generation, who watched the Corps pour its energy into building a MAGTF capable of leading the march north to Bagdad in 2003, Steel Knight 2025 represents a refreshing return to ensuring the Marine Corps remains relevant across the full range of military operations. The narrative described by Laird will feel familiar and new at once. Familiar, because every major shift in our history has been driven from the intersection of strategy, technology, and hard-earned operational experience. New, because the environment we are preparing for — contested logistics, persistent surveillance, cyber and information attacks, compressed decision cycles — demands that we think of ourselves less as a self-contained combined arms team and more as the connective tissue of a wider joint and allied kill web.

Young Marines picking up this book will find more than a description of where we are; they will find a candid account of the problems they are being asked to solve. They will see that their ingenuity at a FARP in the desert; their judgment on a spectrum-crowded net; their willingness to exercise mission command rather than wait for direction, are not side notes to some larger theory of war — they are the theory, in practice. Allies and partners will recognize in these pages a Marine Corps determined to operate as a reliable combat force, able to sense, decide, and enable fires across the coalition battlespace.

During my time in uniform, I saw concepts rise and fall on whether they were disseminated and embraced by Marines at every level - from Lance Corporals manning a radio net on a remote ridge to Colonels in a wing operations center. What *Building the Impact Force* documents is that this transformation is already in their hands. The task of senior leaders is to give them the authorities, the structure, and the resources needed to succeed, and to have the humility to adjust when the *Campaign of Learning* shows us we were wrong.

This book will not settle the debate surrounding Force Design 2030, nor does it pretend to. It does something more useful. It captures, in real time, how the Corps is moving from being very good at managing crises to being able to impose order and advantage in chaos. For a nation that will increasingly rely on forward, resilient, integrated forces to deter and, if necessary, defeat peer adversaries, that is the standard that matters.

Robbin Laird has been chronicling the Corps' journey for decades. With this volume, he offers Marines, our joint teammates, and our allies a clear-eyed view of where we stand and what remains to be done. I commend it to you as both a record and a challenge: a record of how far we have come, and a challenge to every Marine to help build the impact force our nation demands.

CONTENTS

PART THREE
WORKING THE TRANSFORMATION

PROLOGUE

The argument of my 2022 book on the United States Marine Corps was straightforward. The Corps was already well down a transformation path before the current round of controversies about

Force Design 2030 erupted, and that path centered on preparing Marines for the high-end fight and full-spectrum crisis management after the long detour of the Middle East land wars.

In that book, I traced how the Marines, particularly through their aviation community, had begun to reorient toward peer competitors, mobile and expeditionary basing, and tighter integration with the Navy and Air Force well before Afghanistan ended and before most of Washington fully grasped the strategic shock we were living through.

This new book takes that earlier assessment as a point of departure and asks a different, harder question not simply whether the Marine Corps is transforming, but whether it can become what I call an impact force, a force designed from the outset to generate disproportionate effects for the joint and combined team under contested, chaotic conditions?

In other words, the story now is less about "Are the Marines leaving the land wars?" and more about "Can the Marines deliver meaningful, sustained combat and deterrent effect inside an adversary's weapons engagement zone while enabling the larger kill web to function?"

To answer that question, we need first to recall where the Corps actually was in 2021–2022, what my previous work concluded about the transformation path, and where subsequent experience especially exercises like Steel Knight 2025 and the evolution of Marine aviation and assault support has validated, modified, or challenged those conclusions.

In the earlier book, I argued that the 2021 withdrawal from Afghanistan marked more than the end of a campaign. It exposed the bankruptcy of a CENTCOM-dominated, Army-centric way of war that had shaped U.S. force design, budgeting, and operations for two decades. The Marines had become, in large measure, an adjunct to that model, providing infantry, aviation, and logistics to support counterinsurgency, counterterrorism, and nation-building operations in the greater Middle East. Marine aviation operated largely as an enabler for the "ground scheme of maneuver," and even when air and sea power were used creatively, they remained harnessed to a land-centric conception of joint warfighting.

But as I highlighted then, this period also saw the seeds of a quite different Marine Corps. The introduction of the MV-22B Osprey and later the F-35B/C created an aviation enterprise inherently more suited to contested, multi-domain operations than to static, FOB-based counterinsurgency. Marines were among the first to understand that platforms like the Osprey were not mere replacements for legacy helicopters but enablers of new operational concepts for range, speed, and distributed maneuver, even if doctrine and structure lagged behind. At the same time, some senior Air Force and Navy leaders, whom I worked with during the Gates–Obama period, warned that pouring resources into the land wars would erode high-end capabilities, particularly fifth-generation airpower and anti-submarine warfare. Those warnings proved prescient.[*]

By 2020–2021, the strategic environment had changed decisively. Peer competitors Russia and China were shaping military tools for operating below the threshold of large-scale war but with lethal capabilities that could rapidly escalate a crisis. The Middle East template offered little guidance for this world. My conclusion in 2022 was that success in the "strategic shift" would require reworking how joint and allied forces operated together, and that the Marines, because of their size, culture, and air-ground-sea mix, were uniquely positioned at the vortex of that change.

THE MARINES AT THE VORTEX OF CHANGE

That earlier book was built largely from the perspective of Marine operators and commanders in key commands MAWTS-1, 2nd MAW, II MEF, MARFORPAC, and MARSOC. What emerged from those interviews was not a service waiting to be redesigned from above, but one already improvising a way forward under operational pressure.

At MAWTS-1, for example, we saw the Corps' aviation center of excellence acting as a crucible for integrating new platforms and for

---

[*] Robbin Laird, editor, *America, Global Military Competition, and Opportunities Lost: Reflections on the Work of Michael W. Wynne* (Second Line of Defense, 2025).

developing tactics, techniques, and procedures for mobile and expeditionary basing. The Weapons and Tactics Instructors were already grappling with questions that sit at the heart of today's debates How can the Marine Corps most effectively contribute to sea control and sea denial in the Pacific? How do you bring "pieces on the chessboard" inside adversary rings in ways that minimize risk and maximize political and military effect? How do you train to operate from afloat and ashore as part of a distributed naval fight, rather than simply moving troops from ship to shore as a sort of maritime bus service?

The answers were necessarily provisional, but the direction was clear. Mobile basing was treated as a core competence, not a niche. Expeditionary airfields, FARPs, and distributed hubs and spokes were being used to extend reach and complicate adversary targeting, while integrating Marine air with Navy and Air Force strike and C2 assets. The F-35 was already functioning as a key enabler of this shift, not just as a strike platform but as an information and integration node capable of extending the "big blue blanket" of situational awareness and cueing over the maritime battlespace.

Similarly, discussions at II MEF and with 2nd Fleet and JFC Norfolk highlighted the Marines' potential role in North Atlantic defense as part of a distributed, integrated force. The question there was not whether the Corps would be relevant, but how to organize C2, logistics, and air-sea integration so that Marine formations could contribute meaningfully to the 4th Battle of the Atlantic, including in the High North and Nordic regions. Again, the emphasis was on integration with allied airpower and naval forces and on exploiting the F-35 as a multinational, multi-domain integrator.

In the Pacific, my engagements with MARFORPAC underscored a similar pattern. The so-called "distributed laydown" begun earlier in the decade was being reworked in light of INDOPACOM realities, Force Design 2030, and the need to operate inside Chinese weapons engagement zones. Marines in Hawaii, Okinawa, and beyond were focused less on static posture and more on the ability to move, sustain, and protect small but consequential units that could provide sensing, fires, and C2 functions for the joint force. MARSOF 2030, which I examined as a case study, suggested that MARSOC could play an important role as a connector, synchronizer, and small-unit innovator in this environment, contributing to below-threshold operations and to the "inside force" concept that would later become prominent in Force Design debates.[*]

My 2022 conclusion was that the Marines were already partway down a transformation path anchored in three themes

- Shifting from land-war adjunct to full-spectrum crisis management and high-end warfighting.
- Leveraging aviation transformation (Osprey, F-35, digital interoperability) to enable distributed operations and naval integration.
- Experimenting with new C2, logistics, and small-unit models that anticipated later Force Design 2030 controversies.

What I did not yet have, and what this new book provides, is a sustained assessment of how far that path has taken us toward an impact force model and where the gaps remain.

FROM TRANSFORMATION PATH TO IMPACT FORCE

The notion of an "impact force" emerged from my more recent work on exercises like Steel Knight 2025, the evolution of assault support and aviation, and the Marine Corps' own "campaign of learning" under Force Design. The basic shift is conceptual.

---

[*] MARSOF 2030 is Marine Corps Forces Special Operations Command's strategic vision for how Marine Raiders will evolve to remain relevant and effective in a more complex, data-rich, and politically constrained operating environment by the 2030 time frame.

MARSOF 2030 treats 2030 as an "aimpoint" to guide investment, force development, and organizational change beyond the current FYDP, shifting MARSOC from a primarily counterterrorism, needs-based force to one optimized for great-power competition, gray-zone conflict, and multi-domain operations. It frames the future environment around two converging conditions, regional competition and instability, which together drive demand for small, networked, high-cognition SOF elements that can sense, understand, and influence complex systems rather than just conduct episodic raids.

The earlier book was concerned primarily with whether the Marines were escaping the gravitational pull of the land wars and repositioning themselves for Great Power competition. This book is concerned with a more demanding question whether they are building a force that can deliver sustained, disproportionate effects across joint kill webs in the face of contested logistics, degraded communications, and political and operational ambiguity.?[*]

Two key ideas structure this shift.

- The first is the move from a crisis-management force to an impact force. Traditionally, the Corps was optimized to respond rapidly to discrete crises embassy reinforcements, non-combatant evacuations, limited interventions with clear beginnings and ends. Force packages were built around relatively self-contained Marine Air-Ground Task Forces operating under fairly clear political direction. Today's environment demands something different. The boundary between peace and war has eroded, adversaries operate continuously in the gray zone, and the speed of decision-making has increased dramatically as sensors, networks, and algorithms compress timelines. In this world, Marines must be able to generate impact not just by what they shoot, but by how they sense, connect, and enable others to shoot whether those "others" are Navy surface combatants, Air Force bombers, allied F-35s, or unmanned systems.
- The second idea is that of decision-centric, kill-web enabled operations. In my analysis of the 2025 and 2026 Marine Aviation Plans, Project Eagle, and the elevation of Aviation Ground Support as the seventh function of Marine aviation, it is clear that the Corps is intentionally pivoting toward decision advantage as a central objective for the Aviation

---

[*] For an analysis of the kill web concept, see Robbin Laird and Edward Timperlake, *A Maritime Kill Web Force in the Making: Deterrence and Warfighting in the 21st Century* (Amazon: 2022).

Combat Element. Digital interoperability, collaborative combat aircraft, and distributed aviation operations are not simply technical upgrades they are the mechanisms by which Marine aviation becomes a key node in a broader decision-centric network. In practical terms, this means measuring success less by sortie counts or tonnage delivered and more by how much decision space and tempo advantage Marines can create for the joint force.

LESSONS FROM STEEL KNIGHT AND THE CAMPAIGN OF LEARNING

These conceptual moves are not occurring in a vacuum. They are being tested, refined, and sometimes challenged in a series of exercises, experiments, and studies that the Corps rightly casts as a "campaign of learning." Among these, Steel Knight 2025 stands out as a watershed, integrating Force Design 2030 updates, distributed operations, and kill-web experimentation at a scale that provided realistic friction and insight.

From that exercise and related work, several themes emerge that are central to this new book.

- First, the operational logic of distributed, impact-oriented operations is sound but incomplete. Steel Knight showed that Marines can, in fact, operate as distributed nodes that generate targeting data, coordinate fires, and maintain survivability through movement and signature management. Hub–spoke–node constructs, time-stamped nodes with inherent "shelf lives," and integrated strike–assault packages supported by V-22s, CH-53s, and F-35s are not theory they are being practiced. HMLA-267's transition to full digital interoperability, and the way its H-1s now function as C2 and sensing nodes as much as traditional shooters, exemplifies this shift.
- Second, the constraint set is real and severe. Logistics, in particular, remains the limiting factor. Distributed operations

over Pacific distances demand more KC-130Js, MV-22s, CH-53Ks, and maritime autonomous systems than the Corps currently possesses, and contested logistics impose demands that legacy supply chains and sustainment models struggle to meet. Exercises have made clear that you "cannot do this operation with magic," in the words of one commander they require deliberate investments in lift, fuel, predictive maintenance, and aviation ground support, as well as creative use of pre-positioning, joint support, and allied infrastructure.

- Third, command and control and delegation lag behind the technology. The Corps' culture of mission command and maneuver warfare is an asset, but the availability of rich, shared information through data links and networks creates a real tension between centralized oversight and decentralized execution. Steel Knight and related efforts show that operating in an environment where a 30-second transmission can invite rapid retribution demands new patterns of authority, trust, and risk acceptance. The question of how much decision authority to push to small distributed nodes and under what conditions remains unresolved, and it is central to whether the Marines can truly function as an impact force rather than as a collection of well-sensed but tightly-controlled units.

- Fourth, the Corps is willing to adapt in light of evidence, but that adaptation is uneven and contested. The evolution of Force Design 2030 between its original 2020 formulation and the October 2025 update reflects a willingness to reverse or modify decisions on issues such as breaching capabilities, MLR numbers, and the balance between "stand-in" and traditional infantry forces. The retention of 4th Marines as a conventional infantry regiment, and the recognition that some bridging and breaching capabilities divested with tanks must be restored, shows responsiveness to exercise-derived lessons. At the same time, critique from senior retired leaders underscores unresolved debates about over-specialization, amphibious capability, and global crisis response.

THE AGENDA OF THIS BOOK

This book builds on the 2022 assessment but approaches the transformation from a different angle. Rather than asking "Is the Marine Corps ready for the high-end fight?" which assumes fight preparation as the metric of success, I ask "Is the Marine Corps evolving into an effective impact force for an age of chaos?" This reframing reflects the conceptual evolution documented above and focuses analysis on different dimensions of transformation.

The book draws extensively on Steel Knight 2025, which functioned as a campaign laboratory for impact force operations. Unlike previous large-scale exercises designed primarily for readiness assessment, Steel Knight deliberately tested whether Marines could function as forward, resilient, digitally integrated elements of joint kill webs rather than self-contained combined arms teams. The exercise stressed logistics, command and control, and digital interoperability under realistic constraints, revealing both progress and persistent challenges.

The analysis examines three interlocking dimensions of the impact force concept.

- First, aviation as impact multiplier: how Marine aviation is being reconceived as connective tissue for distributed operations rather than a collection of platforms aligned under traditional mission categories.
- Second, kill web integration through digital interoperability: how Marines are learning to contribute to and benefit from networked sensors, shooters, and decision-makers across services and allies.
- Third, force design as impact logic: how the October 2025 Force Design Update and related guidance reflect an organization learning through iteration rather than implementing a fixed blueprint.

Throughout, the emphasis is on understanding transformation as process rather than event. The Marine Corps is not implementing a predetermined design; it is conducting a campaign of learning through field experimentation, war gaming, and operational experience. Failures and friction are features of this approach, not bugs. The question is whether the organization is learning from those failures quickly enough to adapt faster than adversaries can counter.

Looking back at the 2022 analysis, I see both validation and challenge. The fundamental drivers I identified, adversary modernization, technological disruption, and the shift from land wars to maritime competition, remain correct. Marine aviation's leadership role in transformation continues. The challenges of C2, logistics, and cultural change persist.

But the transformation has evolved in directions that go beyond refinement of the path documented in 2022. The shift from crisis management to chaos management, from high-end fight preparation to impact force development, from service distinctiveness to joint enablement, these represent not incremental adjustments but fundamental reconceptualization of what Marines are for in contemporary competition and conflict.

The chapters that follow trace this evolution through operational experience, strategic guidance, and the candid assessments of Marines grappling with transformation in real time. The goal is not to render final judgment on whether Force Design 2030 has succeeded or failed. It is far too early for such conclusions but to understand how the transformation is actually unfolding, what it reveals about modern military adaptation, and what choices lie ahead as the Marine Corps continues its journey from crisis management force to impact force in an age of chaos.

This book therefore does not begin from a blank slate. It starts from the baseline established in my 2022 assessment of the USMC transformation path and uses that as a reference point to examine the next phase the emergence, challenges, and prospects of the Marine Corps as an impact force.

The focus is on practice as much as on doctrine how Marines, sailors, and airmen are actually operating in exercises and deployments, how they are using new platforms and networks, and how they are wrestling with logistics, delegation, and integration problems that no glossy concept paper can resolve.

Throughout, I will return to the core question posed at the outset Can the Marine Corps not only leave the land-war paradigm behind, but build and sustain an impact force capable of making a decisive difference in the kind of contested, fast-moving, and politically ambiguous conflicts that now define Great Power competition?

My 2022 book showed that the Corps had already embarked on a serious transformation path. This book takes up the story at that point and examines whether, and under what conditions, that path can lead to the kind of impact force the nation and its allies now require.

INTRODUCTION

The United States Marine Corps transformation, first chronicled in my 2022 book as a departure from Middle East land wars toward high-end maritime competition, has now reached a pivotal evolution. The Marine Corps is becoming focused on becoming an impact force within the joint force, one that generates disproportionate effects across joint kill webs in contested chaos, rather than merely responding to discrete crises. The question is how to shape its ability to operate as an impact force as the key element and as a component of the joint force.

In my view, the Marine Corps needs capabilities to ensure it can operate as a force delivering key effects even without other elements of the joint force. But like the Ukrainian forces relying on allied ISR kill webs, the Marine Corps will not deploy alone even when it is the lead or corps element.

For decades post-Cold War, Marines excelled as a crisis management force: forward-deployed for embassy reinforcements, evacuations, or humanitarian missions with clear start and end points. That model assumed stability punctuated by emergencies, supported by rehearsed playbooks and self-contained Air-Ground Task Forces.

But adversaries like China erode peace-war boundaries with gray-zone operations, cyber disruptions, and precision strikes, compressing decision cycles beyond traditional responses. The Marine Corps is now adapting into an impact force, persistent, resilient nodes that sense, connect, and enable distant strikes rather than dominate battlespaces alone.

This evolution redefines success. The question is no longer "How fast can a MAGTF execute its mission?" but rather "How much decision advantage do Marines provide the joint force through target discovery, data sharing, and survivable presence?"

These metrics draw from the Campaign of Learning, especially Steel Knight 2025, where I MEF tested distributed operations against peer threats with no permissive buildup, just immediate dispersal, fires integration, and logistics under fire. Steel Knight revealed both strengths (F-35s as network hubs, MV-22/CH-53K logistics spokes) and gaps (contested sustainment, C2 delegation lags). Force Design updates, like retaining 4th Marines and restoring breaching capabilities, show evidence-based iteration prioritizing joint enablement over rigid blueprints.

At the heart of these trade-offs is a simple truth: impact force design does not eliminate risk. It trades one set of risks for another.

The transformation is not about creating an invulnerable force but about aligning vulnerabilities with the realities of contemporary competition and conflict, a force that can absorb friction, adapt under pressure, and continue creating operational dilemmas for adversaries even when parts of the system are degraded or cut off. This demands not only different force structure and technology but also a different mindset about what "success" and "failure" look like in combat.

The chapters that follow trace this evolution along three interlocking dimensions:

Aviation as an Impact Multiplier

Marine aviation is no longer understood primarily as a set of platforms aligned neatly under traditional mission categories like offensive air support or assault support.

As codified in the 2026 Marine Aviation Plan and visible in Steel Knight's scenarios, aviation is being reconceived as the connective tissue of distributed operations. F-35s, MQ-9s, MV-22s, CH-53Ks, and H-1s increasingly function as networked nodes in kill webs, as sensors, gateways, airborne C2 platforms, and logistics enablers, rather than as isolated shooters. The elevation of Aviation Ground Support to a formal "seventh function" reflects recognition that the ability to create and sustain expeditionary hubs, spokes, and nodes, generating fuel, power, maintenance, and basic airfield services from austere, shifting locations, is what makes the entire aviation enterprise relevant to impact force operations.

Kill Web Integration Through Digital Interoperability

The real revolution in Marine Corps warfighting does not lie in any single new munition or airframe but in the effort to connect existing and emerging sensors, shooters, and decision-makers across services and allies into resilient, adaptive networks.

Initiatives such as the modernization of Marine air command and control, the proliferation of digitally interoperable platforms, and the integration of AI-enabled tools all attempt to answer the same question: How do Marines contribute to and benefit from kill webs that must function at machine-speed under attack?

The answers are uneven and incomplete, but the direction of travel is clear: success will depend less on owning striking platforms and more on being able to create, manage, and exploit the data flows that make those platforms effective.

Force Design as Impact Logic

The October 2025 Force Design Update, viewed through the lens of Steel Knight and related experiments, reads less like a static blueprint and more like a record of a learning organization adjusting course in real time.

The decision to shift emphasis from "ship killing" to sensing and enabling; the choice to retain 4th Marines as a reinforced infantry regiment rather than converting it to another Marine Littoral Regiment; the quiet restoration of breach and gap-crossing capabilities that earlier iterations had divested, all reflect a willingness to let evidence override doctrine.

They also highlight unresolved debates about the balance between specialization and flexibility, between Indo-Pacific prioritization and global responsiveness, and between aspirational concepts and industrial base realities.

Running through all three dimensions is a theme that this book treats as decisive: culture and human factors will determine the success or failure of the impact force far more than any single platform or program.

The Marine Corps can field digitally interoperable aircraft, deploy AI-enabled decision aids, and build exquisite kill web architectures, but if its command philosophy and training systems default to centralization under stress, much of that investment will be wasted. The captain commanding a distributed node with access to F-35 feeds, MQ-9 coverage, and joint fires must have the authority and expectation to act within intent, not wait for direction from an overwhelmed headquarters.

Mission command cannot remain an aspirational slogan in doctrine; it must become lived practice in exercises, deployments, and crises. That requires sustained work on education, trust, and the institutional incentives that shape behavior when uncertainty and risk spike.

This is no solved puzzle, critics rightly highlight gaps in contested logistics, C2 resilience, and industrial capacity. But standing still invites irrelevance.

Grounded in field observation and practitioner insights, this book captures a force in motion, trading the familiar risks of crisis management for the harder-to-predict risks of chaos management in an era where adversaries refuse to play by our preferred rules.

PART ONE
THE EMERGENCE OF
THE IMPACT FORCE

Part one of the book establishes what it means for the United States Marine Corps to transition from a crisis-response force to an impact force designed to operate inside chaos, not on the margins of discrete, time-bounded emergencies.

This shift is presented not as a matter of slogans or PowerPoint, but as a hard-earned evolution visible in exercises like Steel Knight 2025, in the October 2025 Force Design Update, and in the 2026 Marine Aviation Plan's re-architecture of aviation, command and control, and logistics around kill web operations.

The opening chapters frame a service grappling with how to generate disproportionate effects for the joint force when decision cycles are compressed, sanctuaries are disappearing, and every node that connects the force also makes it targetable.

The first chapter defines the impact force concept and distinguishes it sharply from both traditional crisis management and earlier "inside force" thinking. For decades, Marines optimized for rapid, scripted responses to discrete contingencies from relatively stable basing, treating crises as episodic deviations from normalcy. Chaos management assumes the opposite: multiple, overlapping threats across domains, blurred lines between peace and war, and an operating environment that is permanently contested. Within this environment, the chapter differentiates between an "inside force" that merely survives inside an adversary's weapons engagement zone and an "impact force" that serves as the sensing, targeting, and command-and-control layer that makes the rest of the joint force far more lethal and survivable. The impact force is measured less by what it kills with its own magazines than by the tempo and quality of effects it enables across a distributed kill web.

From this conceptual baseline, three dimensions are set out along which Marine Corps transformation must be understood: aviation as impact multiplier, kill web integration through digital interoperability, and force design as impact logic.

- Aviation is recast from a collection of fires platforms into the connective tissue of distributed operations, with the F-35, MQ-9, MV-22, CH-53K, and even the H-1 treated as airborne nodes that sense, fuse, and broker information as much as they deliver ordnance.
- The Digital Interoperability for MAGTF and Naval Integration effort, Marine Air Operations Center modernization, and a growing ecosystem of gateways, sensors, and AI/ML tools are presented as the enabling layer that turns platforms into a web.
- Force Design decisions. from retaining 4th Marines instead of converting it to another Marine Littoral Regiment, to restoring breaching capabilities, to pivoting from a shooter-centric to a sensor-centric view of forward units, are interpreted as evidence that the service is rethinking its theory of value in terms of impact, not organic firepower alone.

This evolution is described as creating as many vulnerabilities as strengths, and technology is shown as unable to substitute for culture. Digital interoperability and distributed operations promise faster decisions and greater survivability, but they also generate dense electromagnetic signatures, new dependencies on networks and data, and acute stress on logistics and training systems. AI/ML-supported pattern recognition and predictive maintenance may enhance awareness and readiness, yet they introduce new failure modes and potential single points of systemic weakness.

Above all, whether the Marine Corps successfully becomes an impact force is portrayed as depending on institutional ability to embed mission command in a truly contested, high-velocity environment, push authority to junior leaders who possess unprecedented situational awareness, and resist the instinct to recentralize control whenever pressure mounts.

Subsequent chapters in part one take the reader into the campaign laboratory where this theory is being tested: Steel Knight 2025. Rather than a traditional readiness check, Steel Knight is presented as the first full-scale rehearsal of the Marine Corps operating as an impact force under chaos conditions, with I MEF executing a linked campaign across Southern California and the broader Southwest from the opening moments of a simulated crisis. The exercise abandons linear force closure in favor of distributed entry operations, assuming that there will be no secure buildup window and no sanctuary from long-range fires and persistent ISR. It forces commanders to fight dispersed from the start, constantly move nodes to survive, and rely on networked integration rather than massed fires to generate effects.

Within this campaign, Steel Knight ties experimentation directly to real operational stakes by serving as the certification venue for the regiment deploying as Marine Rotational Force–Darwin. Embassy reinforcement under pressure, large-scale noncombatant evacuations, distributed fires across extended ranges, and contested logistics scenarios are deliberately sequenced to expose interdependencies and compound friction.

The exercise reveals that nodes are often too "heavy," with logistics, protection, and command requirements creating signatures that sophisticated adversaries can rapidly target, leading mentors such as Lieutenant General (Ret.) Robert Hedelund to argue for putting "timestamps" on nodes and treating them as temporary constructs with inherent shelf lives rather than static assets.

It also shows that logistics, especially the dependence on enablers like Air Force C-130s, is the binding constraint for impact forces; without credible, survivable sustainment, even the most elegant kill web quickly collapses.

Digital interoperability is highlighted as both revolutionary and dangerous. Units such as HMLA-267 demonstrate how legacy platforms become critical command-and-control and data nodes once they are fully digitally interoperable, pulling and pushing information across the MAGTF and joint force.

At the same time, every transmission can attract lethal attention within minutes, forcing a new operational calculus about emission control, bandwidth management, and the tradeoff between being connected and being survivable.

Command and control emerges as the main cultural friction point: technical architectures now give captains and sergeants access to sensor data and strike options once reserved for generals, but institutional habits, approval chains, and risk aversion often lag behind, threatening to turn connectivity into a bottleneck rather than an advantage.

Throughout part one, the war in Ukraine serves as a cautionary benchmark rather than a template. What Steel Knight participants draw from Ukraine is not a specific tactic but the premium on adaptation speed, on the capacity to integrate new technologies, tactics, and commercial tools faster than an adversary. The chapter warns that this is not inherently a Western advantage and that any assumption that U.S. forces will automatically adapt faster is dangerously complacent. For the Marine Corps, the challenge is to institutionalize an experimental mindset, treat large-scale exercises as genuine laboratories rather than scripted validations, and be willing to adjust concepts and structures when field evidence contradicts approved ideas.

Part one concludes by arguing for practitioner-grounded assessment of what is actually emerging, why it matters, and where critical risks remain, rather than uncritical endorsement or reflexive dismissal of transformation.

The impact force is portrayed as real and already taking shape in how Marines design exercises, adjust force structure, and recast aviation and logistics as enablers of joint kill webs.

Whether it becomes decisive is left contingent on near-term choices about culture, authorities, logistics investment, and the willingness to let empirical campaign laboratories, not concepts alone, drive the next phase of Marine Corps evolution.

CHAPTER 1
DEFINING THE
IMPACT FORCE

The United States Marine Corps is in the midst of a fundamental transformation, one that extends far beyond new platforms or organizational charts. What emerged during Steel Knight 2025 was evidence of something more profound: the evolution from a crisis response force designed to manage discrete contingencies into what I term an impact force optimized to generate decisive effects within joint kill webs operating under conditions of chaos rather than crisis.

This distinction matters. A crisis management force prepares for known contingencies with established playbooks. An impact force operates when playbooks fail, when multiple threats converge across domains, and when the tempo of operations outpaces traditional decision cycles. It doesn't just respond to problems. It creates disproportionate operational effects that shape the competitive environment.

The October 2025 Force Design Update, the 2026 Marine Aviation Plan, and the operational patterns visible during Steel Knight 2025 collectively reveal a Marine Corps evolving, sometimes deliberately, sometimes through adaptive pressure, into precisely this kind of force. This book examines that evolution, tests its coherence against operational reality, and identifies where the concept remains fragile or contested.

CRISIS MANAGEMENT VERSUS CHAOS MANAGEMENT

For decades, the Marine Corps organized around crisis management. The logic was straightforward: position expeditionary forces forward, maintain readiness for rapid response, execute well-rehearsed missions, embassy reinforcements, noncombatant evacuations, humanitarian assistance. The operating environment was assumed to be relatively stable between crises. When crises erupted, they were treated as discrete events with clear beginnings and endings.

Chaos management operates under different assumptions. Multiple threats emerge simultaneously across geographic and functional domains. Adversaries operate continuously in gray zones below traditional thresholds of armed conflict. The distinction between peace and war blurs. Technological change compresses decision timelines. The competitive environment itself becomes the mission space rather than a series of episodic crises punctuating periods of normalcy.

This shift from crisis to chaos creates fundamentally different force design requirements:

Crisis Management Force	Chaos Management Force
Optimized for discrete contingencies	Designed for persistent competition
Linear escalation assumed	Non-linear, multi-domain threats
Response to problems	Generation of disproportionate effects
Platform-centric capabilities	Network-enabled integration
Centralized decision authority	Distributed command and control
Known adversaries, known capabilities	Adaptive threats across spectrum
Geography defines deployment	Effects define employment

The Marine Corps did not consciously design Force Design 2030 around chaos management theory. But the operational logic emerging from exercises like Steel Knight 2025 reflects precisely the kind of adaptive, distributed, effects-focused approach that chaos management demands.

INSIDE FORCE VERSUS IMPACT FORCE

Within the chaos management framework, another critical distinction emerges: the difference between an inside force and an impact force. These terms, while related to geography, are fundamentally about operational function rather than physical location.

An inside force positions itself within an adversary's weapons engagement zones to complicate targeting, impose costs, and enable joint force operations. The concept drove early Force Design thinking, Marine Littoral Regiments equipped with anti-ship missiles operating from island chains to deny sea control.

An impact force generates effects disproportionate to its size by serving as a sensor, targeting, and command-and-control enabler for joint kill webs. It doesn't just survive inside threat rings; it makes everything else the joint force brings to the fight vastly more lethal and survivable.

This evolution represents the single most important shift in Force Design thinking between 2020 and 2025. As Brigadier General Christopher Haar, Assistant Deputy Commandant for Combat Development and Integration (CD&I) at Headquarters Marine Corps, articulated, the original concept was "all about killing ships for the stand-in force." Through the Campaign of Learning, leaders recognized "we're better as the JTAC for the joint force, using our sensors, using our eyes, killing occasionally, but using our time and our capabilities to enable the rest of the joint force."[*]

The distinction matters:

---

[*] Hope Hodge Seck, "Marine Corps Master Plan Evolves in New Update," *National Defense*, February 11, 2026, https://www.nationaldefensemagazine.org/articles/2026/2/11/marine-corps-master-plan-evolves-in-new-update.

Inside Force	Impact Force
Geographic positioning	Operational effect generation
Direct-fire emphasis	Sensor-shooter integration
Platform survival	Network viability
Self-sufficient operations	Joint force enablement
Anti-access focus	Multi-domain connection
Unit-level lethality	System-level decision advantage

An inside force can become an impact force, but the transformation requires different capabilities, different command relationships, and fundamentally different metrics of success. You measure an inside force by its ability to survive and strike. You measure an impact force by how much it amplifies the lethality and decision speed of forces that may be hundreds or thousands of miles away.

THREE DIMENSIONS OF IMPACT FORCE EVOLUTION

This book examines the Marine Corps' evolution toward becoming an impact force through three interconnected dimensions:

1. Aviation as Impact Multiplier

Marine aviation is evolving from a fires platform into the connective tissue of distributed operations. The 2026 Marine Aviation Plan's emphasis on Digital Aviation Operations (DAO), the Deputy Commandant for Aviation (DCAO) construct, and Aviation Ground Support (AGS) as a seventh functional capability reflects this shift. Platforms like the F-35 and MQ-9 increasingly function as airborne nodes in kill webs rather than traditional shooters. The H-1 is becoming a rotary-wing command and control platform. Even logistics platforms like the MV-22 and CH-53K enable impact by sustaining distributed nodes that generate targeting data.

2. Kill Web Integration Through Digital Interoperability

The real revolution isn't new weapons: it's connecting existing sensors and shooters with the speed and reliability that chaos demands. The Digital Interoperability for MAGTF and Naval Integration (DI/MANGL) effort, Marine Air Operations Center modernization, G/ATOR radar networking, and experiments with AI/ML for pattern recognition all reflect a service trying to solve the fundamental problem: how do you generate decision advantage when the tempo of operations exceeds human processing speed?*

3. Force Design as Impact Logic

The October 2025 Force Design Update's key decisions—retaining 4th Marines rather than converting it to a third MLR, restoring breaching capabilities, pivoting from "shooter" to "sensor" emphasis—make sense when viewed through an impact force lens. The Corps is preserving flexibility while building specialized capabilities that enable joint force effects generation. The debate over "China-only" versus globally responsive forces reflects unresolved tension about what kind of impact the Corps should generate and where.

THE ARGUMENT OF THIS BOOK

This analysis develops three interconnected arguments:

- First: The Marine Corps is evolving into an impact force optimized for enabling joint kill webs in chaos management conditions. This evolution is visible in Steel Knight 2025 operational patterns, Force Design adjustments, and the 2026 Aviation Plan architecture. It is real, not rhetorical.
- Second: This evolution creates new vulnerabilities even as it

---

* In USMC usage, "DI/MANGL" refers to the combined concept/program Digital Interoperability / MAGTF Agile Network Gateway Link.

generates new capabilities. Digital interoperability makes you faster—it also makes you targetable. Distributed operations complicate adversary targeting—they also stress command and control, logistics, and training. AI/ML enables pattern recognition beyond human capacity. It creates new dependencies and failure modes. Impact force design trades one set of risks for another; it doesn't eliminate risk.

- Third: Culture and human factors will determine whether the Marine Corps successfully becomes an impact force more than technology or platforms. Mission command philosophy must evolve faster than weapons systems. Young leaders need authority commensurate with their technical capabilities. The institutional tendency toward centralization under stress must be overcome or the entire concept fails.

CHAPTER ARCHITECTURE: WHAT DOES IMPACT FORCE LOOK LIKE?

Each subsequent chapter in this part of the book asks a single organizing question: "What does an impact force look like here?"

Chapter 2 examines Steel Knight 2025 as the first full-scale rehearsal of impact force operations, analyzing how the exercise's campaign design, embassy reinforcement, noncombatant evacuation, distributed fires, contested logistics, and MRF-D certification—tested the concepts emerging from Force Design.

Chapter 3 explores Force Design updates through an impact force lens, examining why the shooter-to-sensor pivot, the decision to retain 4th Marines, and restored breaching capabilities make operational sense when the goal is enabling joint kill webs rather than operating independently.

Chapter 4 analyzes Marine aviation's transformation into an impact multiplier through Project Eagle's phasing (Fight Tonight / Bridge / Future Fight), the DAO/DCAO command relationship evolution, and AGS as a seventh functional capability.

Chapter 5 dissects the kill web integration challenge, examining digital interoperability efforts, command and control modernization, air defense integration, and the hub-versus-node debate that defines distributed operations viability.

Chapter 6 provides platform case studies, analyzing how the F-35, MQ-9, MV-22, CH-53K, and H-1 contribute to impact force operations rather than simply describing their technical capabilities.

Chapter 7 examines sustainability challenges, connecting Steel Knight logistics stresses with AI/ML predictive maintenance, dynamic supply concepts, and the industrial base realities that constrain what's operationally possible.

Chapter 8 addresses command culture, mission command philosophy, human factors in distributed operations, and the training methodologies required to prepare leaders for decision-making under chaos conditions.

This analysis derives from field research conducted during Steel Knight 2025 and decades of engagement with military transformation across three continents.

I am biased toward practitioners over theorists, field research over Washington briefings, and honest assessment of what's actually working versus institutional narratives about what should be working. I have watched too many military transformations fail because analysts were unwilling to report what they observed when it contradicted approved concepts.*

The Marine Corps deserves better than cheerleading. It also deserves better than reflexive criticism that ignores how profoundly the competitive environment has changed. This book attempts to provide rigorous analysis of what's emerging, why it matters, and where critical risks remain.

The impact force is real. Whether it proves decisive depends on choices the Marine Corps and the joint force it serves makes in the years ahead.

---

* My 2026 book looks at four decades of examination of military transformation and lessons learned. *Lessons in Military Transformation: From the RMA to the Drone Wars.*

CHAPTER 2

IMPACT FORCE IN PRACTICE: STEEL KNIGHT 25 AS CAMPAIGN LABORATORY

Steel Knight 2025 wasn't a readiness check. It was the first full-scale dress rehearsal of the Marine Corps operating as an impact force under chaos management conditions.

Conducted by I Marine Expeditionary Force across Southern California and the greater Southwest from December 1-14, 2025, the exercise abandoned the traditional model of scripted, single-phase battles in favor of something more demanding: a campaign of linked problems requiring distributed operations from the opening moments.

What made Steel Knight 2025 significant wasn't the platforms involved or the number of participants. What mattered was the operating logic it tested. I MEF attempted to validate whether it could function as a forward, resilient, digitally integrated force designed to generate disproportionate effects, not through mass, but through how it sensed, decided, and enabled the broader joint force.

The exercise's campaign design deliberately sequenced scenarios that stress-tested the impact force concept: embassy reinforcement under pressure, noncombatant evacuation operations, distributed fires across extended ranges, and sustained logistics in contested environments. Layered onto this was the operational imperative of certifying the regiment deploying as Marine Rotational Force–Darwin (MRF-D), tying experimentation directly to forward presence requirements in the Indo-Pacific.

This chapter examines what Steel Knight 2025 revealed about the Marine Corps' evolution toward becoming an impact force. It analyzes the exercise's design architecture, the operational concepts tested, the problems exposed, and the lessons learned that shaped subsequent Force Design adjustments.

In part two of the book are the interviews which I conducted during the exercise when I visited 3rd Marine Air Wing. This chapter highlights some of the conclusions from those interviews, but the details provided in the interviews are a major input to the process of shaping the core analysis of this book.

FROM FORCE CLOSURE TO DISTRIBUTED ENTRY OPERATIONS

The design architecture of Steel Knight 2025 represented a fundamental break with how the Marine Corps traditionally trained at scale. Previous large exercises assumed linear force buildup: forces would deploy into theater, establish secure bases, aggregate combat power, and then engage the adversary. Steel Knight 2025 inverted that logic. It required I MEF to fight as a distributed, networked stand-in force from the moment the crisis began.

This doctrinal departure reflects strategic reality in the Indo-Pacific. There may be no time for methodical force closure. There may be no sanctuary from which to aggregate capability. Adversary long-range strike and persistent surveillance mean that concentration invites targeting. The exercise forced commanders to operate under the assumption that forces must be dispersed from the start, must move constantly to survive, and must generate effects through networked integration rather than massed fires.

The shift from force closure to what participants called "distributed entry operations" requires rethinking everything from task organization to logistics to command relationships. Traditional amphibious doctrine assumed relatively permissive environments once forces were ashore. Steel Knight tested whether Marines could establish, sustain, and fight from distributed positions when every logistics node is targetable, every communications link is contested, and every operational movement creates signatures that adversaries can exploit.

This operational framework mirrors the campaign of learning that characterized interwar Fleet Landing Exercises, where the Marine Corps developed amphibious doctrine through iterative experimentation rather than abstract theorizing. Steel Knight 2025 serves the same function for the stand-in force and impact force concepts: translating strategic guidance into operational practice through repeated rehearsal under realistic constraints.

THE MRF-D CERTIFICATION: MAKING TRAINING OPERATIONALLY RELEVANT

Steel Knight 2025 served dual functions: MEF-level training event and formal certification venue for the regiment deploying to Australia as MRF-D. This integration transformed the exercise from theoretical experimentation into operational validation with real stakes. The unit being certified would deploy forward within months. The problems they practiced, distributed operations, contested logistics, embassy reinforcement, NEO, represented scenarios they might actually face in the Indo-Pacific.

The MRF-D certification requirement imposed operational discipline on the exercise design. Scenarios couldn't be artificial or optimistic. They had to reflect genuine contingencies that partner nations worry about. The presence of international observers from Australia, Japan, and regional allies reinforced this imperative. These observers used the exercise as a window into how I MEF trains for high-end conflict and crisis response in contested maritime regions.

This coalition-building dimension shouldn't be underestimated. In an era where strategic competition is as much about maintaining alliances as combat capabilities, exercises like Steel Knight serve theater-shaping functions. They demonstrate U.S. operational concepts to partners and create opportunities for professional military education and relationship-building.

CAMPAIGN ARCHITECTURE: LINKED SCENARIOS AS OPERATIONAL STRESS TEST

The exercise's campaign architecture represented its most significant innovation. Rather than executing isolated scenarios, Steel Knight linked problem sets in ways that compounded friction and revealed interdependencies.

Embassy Reinforcement Under Pressure

The first scenario tested rapid crisis response when an embassy comes under threat in a deteriorating political situation. This wasn't the permissive reinforcement the Marine Corps practiced for decades. The scenario assumed hostile forces with advanced air defenses, anti-ship capabilities, and cyber/electronic warfare capacity.

Noncombatant Evacuation Operations at Scale

The NEO scenario tested whether the Marine Corps could execute large-scale evacuations when airfields are threatened, sea lanes are contested, and the operational timeline is compressed. Previous NEO doctrine assumed Marines could establish secure perimeters around airports or ports and methodically process evacuees.

Distributed Fires and Maneuver: Testing the Kill Web

The distributed fires scenario represented the most demanding test of whether the Marine Corps could function as an impact force. Forces dispersed across hundreds of miles had to discover, characterize, and engage targets while simultaneously managing their own survivability against a peer adversary with sophisticated ISR and long-range strike.

Node Weight and Signature Management

Nodes proved "a little bit too heavy," as one exercise participant noted. The equipment required to establish, sustain, and protect a missile-firing position created signatures that sophisticated adversaries could detect and target. This led to the critical insight from Lieutenant General (Retired) Robert Hedelund who participated as a mentor in the exercise: put timestamps on nodes.

Rather than nodes requesting extraction when threatened, they should operate with inherent shelf lives, measured in hours or days depending on mission. Forces that inserted one node should already be en route to establish the next while the current node completes its mission and prepares to displace. Constant movement becomes essential to survival and effectiveness.

Contested Logistics: The Constraining Enabler

Throughout Steel Knight 2025, logistics emerged as both the critical enabler and potential point of failure for impact force operations. Distributed operations exponentially increase logistics complexity. Every dispersed site requires fuel, water, ammunition, maintenance support, and communications infrastructure. When those sites move constantly to avoid targeting, sustaining them becomes even more difficult.

The exercise revealed dependency on capabilities the Marine Corps doesn't control. Air Force C-130s proved crucial for without them, participants acknowledged, the exercise would have looked dramatically different. One commander's assessment captured the problem: "You cannot do this operation with magic. You might want to, but it's not going to happen. No voodoo here."

Digital Interoperability:
Revolution or Vulnerability?

The exercise revealed that digital interoperability represents perhaps the most significant revolution in military aviation since the jet engine —and also creates new vulnerabilities that adversaries will exploit.

Marine Light Attack Helicopter Squadron 267 (HMLA-267) emerged as the proof of concept for digital interoperability's operational impact. As the first Marine Corps H-1 squadron to achieve full digital interoperability across a complete unit of employment, HMLA-267 demonstrated how platforms sometimes dismissed as vulnerable or obsolete evolve into critical command-and-control nodes for distributed operations.

Command and Control:
The Cultural Friction Point

The exercise demonstrated that distributed operations demand unprecedented delegation of decision authority but institutional culture hasn't fully adapted to that reality.

Modern C5ISR and networking mean that relatively small Marine units possess situational awareness and strike capabilities that once required much larger formations. A Marine captain commanding a distributed node has access to sensor data from F-35s, MQ-9s, space-based ISR, and allied systems. That captain can coordinate fires from naval surface ships, Air Force bombers, and Army artillery. The technical capability exists.

But capability without clear tasking and appropriate authority becomes wasted potential. If that captain must request approval for every action through channels that may be degraded or delayed, the technical advantage disappears. Adversaries operating inside our decision cycle will act while we coordinate.

THE UKRAINIAN PRECEDENT: SPEED OF ADAPTATION

Steel Knight 2025 repeatedly referenced the war in Ukraine but not as a tactical template for Marine Corps operations. The critical lesson participants drew centered on adaptation speed.

Ukrainian forces demonstrated an ability to rapidly integrate commercial drones, modify tactics, and create new operational approaches faster than their adversary. This pattern of rapid battlefield learning cannot be assumed to be a Western advantage. Adversaries globally demonstrate similar capabilities for fast-cycle adaptation.

WHAT STEEL KNIGHT REVEALED ABOUT IMPACT FORCE READINESS

By the time I MEF completed Steel Knight 2025, several realities about the Marine Corps' progress toward becoming an impact force had clarified:

- First: The concept is operationally coherent. Distributed Marines feeding targeting data into joint kill webs while maintaining survivability through constant movement isn't theoretical abstraction. It can work. Steel Knight demonstrated units successfully operating across extended ranges, discovering and characterizing targets, cueing joint fires, and repositioning before adversary fires arrived.
- Second: Capability gaps remain significant. The Marine Corps doesn't have enough platforms, enough digital interoperability, or enough logistics capacity to sustain impact force operations at scale without joint support.
- Third: Logistics is the binding constraint. All the sensors, all the digital interoperability, all the delegated authorities mean nothing if dispersed forces can't be sustained under contested conditions.
- Fourth: Culture lags technology. The Marine Corps has platforms and networks that enable impact force operations.

It doesn't yet have command cultures, training methodologies, and organizational structures fully adapted to exploit those capabilities.

- Fifth: The concept is fragile. Take away Air Force logistics support, degrade satellite communications beyond current planning assumptions, or introduce adversary electronic attack more sophisticated than exercise scenarios assumed, and the impact force concept faces operational limitations.

CONCLUSION: FROM CAMPAIGN LABORATORY TO OPERATIONAL FORCE

Steel Knight 2025 represented essential progress toward transforming the Marine Corps into an impact force. It validated that distributed operations generating disproportionate effects through kill web integration can work. It exposed where capabilities, doctrine, and culture need further development. It provided empirical evidence that shaped the October 2025 Force Design Update's key decisions.

But Steel Knight also revealed how far the Marine Corps still must go. The capability gaps in platforms, the logistics vulnerabilities, the unresolved tensions around command authorities, and the cultural friction points around delegation all emerged clearly. Those problems won't solve themselves through additional exercises.

What matters now is maintaining that intellectual humility and operational pragmatism. Steel Knight 2025 was one iteration in a continuous process of learning, adapting, and refining the force.

The question isn't whether the Marine Corps has perfected the impact force concept. The question is whether it can maintain the experimental mindset, the willingness to adjust based on evidence, and the operational rigor that Steel Knight 2025 demonstrated. Campaign laboratories only generate value if institutions act on what they reveal.

IMPACT ON FORCE DESIGN: FROM SHIP KILLERS TO JOINT JTACS

The October 2025 Force Design Update captures a major recalibration in how the Marine Corps frames its operational contribution: less about being a boutique "ship-killing" force inside the threat ring, and more about being the forward sensing, targeting, and integration layer that makes the joint force more lethal and faster.[1]

What looks, at first glance, like a refinement in language is better understood as a shift in underlying theory of value: Marines create disproportionate effects when they expand the reach, speed, and resilience of joint kill webs rather than when they focus primarily on organic magazine depth.

Brigadier General Christopher Haar, the Marine Corps Warfighting Lab commander, put the logic bluntly in an interview with Defense One: the "original document was all about killing ships for the stand-in force," but planners later realized "we're better as the JTAC for the joint force, using our sensors, using our eyes, killing occasionally, but using our time and our capabilities to enable the rest of the joint force."[*]

That single statement is not a tactical preference; it's a strategic

____

* Stew Magnuson, "Marine Corps Master Plan Evolves in New Update," *National*

admission about comparative advantage in a modern fight defined by long-range precision, pervasive ISR, and the contest for decision advantage.

FROM "INSIDE" TO "IMPACT"

Early Force Design thinking leaned heavily on position, getting Marines "inside" adversary anti-access/area denial systems to complicate targeting, impose costs, and create sea-denial dilemmas, often with organic anti-ship fires as the signature capability. That approach was coherent: forward posture matters, and in littoral geography the ability to "seize and hold key maritime terrain" can shape naval maneuver and deny an adversary freedom of action.

The October 2025 update does not discard this logic; it reframes what Marines primarily do once they are forward: "sense and shape the operating environment, and close kill webs in support of fleet maneuver and joint campaigns."

The operational change implied here is decisive. If the Marine Corps measures effectiveness mainly by what a stand-in force can kill with its own weapons, then the force design conversation stays dominated by launcher counts, reload, survivability of firing positions, and the inevitable arithmetic of limited magazines.

If, instead, effectiveness is measured by how much joint lethality Marines can enable, how many targets they can find, fix, track, characterize, and pass to shooters with deeper magazines, then the Marine Corps naturally prizes sensing, signature management, C2 resilience, and the human/technical skills of targeting and integration.

The Force Design Update explicitly validates this "impact" framing by describing Marines as "forward sensors and shooters, networked to enable kill webs in contested environments," and by emphasizing C2 modernization so Marines can "sense, make sense, and hold targets at risk with speed and resilience."

Put differently: the Marine Corps is positioning itself less as a self-

---

Defense, February 11, 2026, https://www.nationaldefensemagazine.org/articles/2026/2/11/marine-corps-master-plan-evolves-in-new-update.

contained strike complex and more as an indispensable forward node in a joint system designed to compress time from sensing to effects.

WHY THE SHIFT HAPPENED

The document itself points to an empirical engine behind the shift: a "continuous Campaign of Learning tested in wargames, refined in exercises, and proven in real-world operations."

It also highlights how recent major exercises are being used specifically to "create joint and combined kill webs by linking sensors to shooters and fusing operations and intelligence at every participating echelon," while pushing processing and sharing tools forward "so Marines can ingest, process, and share what they sense with the Joint Force."

This transition is a strategic transformation driven by operational learning rather than an abstract concept swap. The internal logic is straightforward: if the contemporary battlefield is saturated with "drones, long-range precision fires, cyber effects, and electronic warfare," and if combat "is unforgiving" with victory going to the side that "adapts faster," then the force that can persist forward, manage signatures, and keep a resilient target-quality picture flowing to the broader joint force will routinely generate more operational advantage than a force optimized mainly for episodic organic shots.

This is where the "JTAC for the joint force" metaphor is useful, not as branding, but as a discipline. JTAC value is not that the controller personally drops the bomb. It's that the controller creates the conditions, identification, deconfliction, timing, terminal control, and confidence in the target, that allow the right shooter to deliver effects at the right moment.

In a maritime-littoral fight, that enabling function scales: one small forward element that survives, sees, correlates multiple feeds, and shares track-quality data can multiply the effectiveness of ships, aircraft, and allied fires far beyond what that element could achieve with its own limited magazines.

WHAT FORCE DESIGN "IMPACT" LOOKS LIKE

The October 2025 update is explicit that modernization is meant to produce forward forces "optimized to operate in the littorals," able to "sense and shape the operating environment, and close kill webs," while also remaining a "globally responsive, lethal, and resilient combined-arms naval expeditionary force."

That's a deliberate balancing act: Marines must still deliver lethal effects, but the center of gravity is shifting toward being a persistent, low-signature, interoperable node that can extend the operational reach of joint commanders.

Several concrete elements in the update map to this emerging impact-force logic:

- Enabling joint kill webs as a core mission. The update does not treat joint integration as a supporting task; it frames Marines as an enabling layer: "Marines enable naval, Joint, and combined kill webs" and provide "forward sensors and shooters." The focus on C2 modernization to "sense, make sense" underscores that the fight is as much about information advantage and resilient connectivity as it is about platform lethality.
- Exercises as integration laboratories. The update highlights a run of exercises (including Steel Knight and others) as vehicles for linking sensors to shooters, integrating national-to-tactical feeds, and hardening the ability of "highly mobile, dispersed units" to function in denied and degraded conditions. That emphasis is consistent with an impact-force approach in which the force's job is to generate and sustain a joint-quality picture and targeting solution forward, under pressure, over time.
- Sensors, networks, and air command/control modernization. The document spotlights air and sensing systems that feed kill webs (e.g., TPS-80 G/ATOR as a sensor feeding joint kill webs) and a broader push to modernize

command-and-control as a "system of systems." It also describes specific initiatives like Project Dynamis, aimed at accelerating contested-environment C2 and targeting, with the goal of connecting "every sensor to every shooter within the Naval Operational Architecture."

- Mobility and logistics as the price of distribution. The update's discussion of littoral maneuver platforms and logistics in contested environments reinforces that distributed operations are not "free". The force must be able to move and sustain itself while being targeted. The text's focus on logistics modernization so distributed forces can "persist in contested environments and remain combat effective" is a practical prerequisite for any impact-force concept.

This also clarifies a subtle but important point: impact-force logic does not require the Marine Corps to be less lethal. It requires the Marine Corps to be lethal in a way that is jointly composable, able to plug into naval and joint architectures, and to create decision advantage by moving targeting-quality information faster than the enemy can disrupt it.

COURSE CORRECTIONS
AND WHAT THEY SIGNAL

The October 2025 update includes visible and quiet "walk-backs" that indicate genuine learning, adjustments in force structure and capability priorities that better align the Corps with operational evidence.

The Force Design Update's decision to retain "4th Marine Regiment... as a reinforced Marine Infantry Regiment" after determining (through the Campaign of Learning) that "two MLRs and one reinforced Marine Infantry Regiment in III MEF is the optimal force composition" is a clear example of this adaptive approach. That decision suggests the Marine Corps is managing risk across mission sets: retaining a conventional infantry structure while still fielding specialized littoral regiments optimized for distributed operations and joint kill-web integration.

Similarly, the update's note that combat engineers are "exploring gap crossing and obstacle breaching capabilities to ensure our forces can close with and destroy the enemy" highlights an operational reality. Even a force oriented on sensing and enabling must still maneuver, and maneuver in contested littorals will confront deliberate obstacles. Force Design is increasingly evidence-driven and is less about doctrinal elegance and more about what repeated exercises and operational constraints reveal.

The strength of Force Design is measured in execution. Marine Expeditionary Forces are not only integrating new capabilities into daily operations, they are proving them in large-scale exercises with allies and partners across the Indo-Pacific, Europe, and North America. These events sharpen readiness, validate emerging concepts, and accelerate the transition of new systems from experimentation to operational use. The Marine Corps has demonstrated its capability to operate in contested environments through a series of complex, multinational exercises that test the full range of Force Design concepts.

BALIKATAN 25. I Marine Expeditionary Force executed a Joint Task Force certification that incorporated live forces and events into a live, virtual, and constructive exercise. This exercise spanned multiple domains to experiment with deploying low-signature, lightweight naval expeditionary formations to test sensing capabilities and rehearse securing and defending key maritime terrain. This event helped to enhance interoperability between the Marine Corps, Japan Self-Defense Forces, and the Armed Forces of the Philippines. For the first time, 3d Marine Littoral Regiment integrated the Navy-Marine Expeditionary Ship Interdiction System, Marine Air Defense Integrated System, and AN/TPS-80 Ground/Air Task-Oriented Radar into this exercise, all within the First Island Chain. The successful integration of these systems demonstrated the Marine Corps' ability to project power and deny adversary freedom of action in the most strategically consequential waters in the world.

KAMANDAG 9 AND TALISMAN SABRE 25. I MEF and III MEF participation in KAMANDAG 9 and TALISMAN SABRE 25 showcased multimodal and mobile command and control and synchronized precision fires across contested terrain. This effort advanced the Marine Corps' ability to fuse intelligence and operations for faster and more precise target engagement. These exercises also improved combat readiness and strengthened alliances between the United States, Japan, Australia, Republic of Korea, and United Kingdom Armed Forces. The exercises validated the concept that distributed forces, properly networked and equipped with precision fires, can hold adversary forces at risk across vast maritime operating areas while remaining difficult to detect and target.

RESOLUTE DRAGON 25. III MEF and the Japan Self-Defense Forces rehearsed Expeditionary Advanced Basing Operations and cross-domain operations to close kill webs in defense of key maritime terrain. RESOLUTE DRAGON 25 is III MEF's premier bilateral exercise with the JSDF across Japan, including the Southwest Islands. The exercise demonstrated how forward-positioned Marine forces, operating from expeditionary locations with limited infrastructure, can contribute to the defense of critical maritime terrain. Marines practiced rapid repositioning, signature management, and the integration of long-range fires into joint targeting networks—all essential capabilities for operating in contested littoral environments.

ATLANTIC ALLIANCE 25. II MEF and 2d Fleet exercised distributed littoral warfare in a contested environment, advancing amphibious capabilities in support of naval maneuver. Marines employed robust, integrated all-domain awareness capabilities and Expeditionary Advanced Basing Operations alongside Dutch, Canadian, and United Kingdom forces both afloat and ashore. This event was the largest amphibious exercise in the Western Atlantic in over a decade and demonstrated that the concepts being refined in the Pacific theater are globally applicable. The exercise validated the Marine Corps' ability to operate as part of NATO and coalition forces, contributing unique capabilities that enable alliance objectives.

NORTHERN EDGE, STEEL KNIGHT, BOLD QUEST, ISLAND MARAUDER, AND PROJECT CONVERGENCE. Participation in these exercises advanced the USMC ability to create joint and combined kill webs by linking sensors to shooters and fusing operations and intelligence at every participating echelon. These events demonstrated resilient networks and processing tools that integrate national, commercial, theater, and tactical feeds to strengthen maritime domain awareness at the tactical edge. Prototypes in multiple form factors now support highly mobile, dispersed units in denied and degraded environments. The USMC is pushing these capabilities forward so Marines can ingest, process, and share what they sense with the Joint Force, enabling faster decision cycles and more precise targeting across all domains.

These exercises represent more than training events. They are proof-of-concept demonstrations that Force Design is operationally viable. The ability to conduct distributed operations, integrate advanced sensing and fires capabilities, maintain command and control in contested electromagnetic environments, and sustain forces forward without reliance on fixed infrastructure has been tested and validated.

Moreover, these exercises have strengthened critical alliances and partnerships, demonstrating that the Marine Corps' transformation enhances not only U.S. capabilities but also the collective strength of democratic nations committed to a free and open Indo-Pacific.

In short, the Marine Corps is not abandoning the idea of operating forward inside contested spaces. It is refining what "success" means when doing so. The shift is toward a force whose primary product is operational impact, measured in targets exposed, kill webs closed, joint fires enabled, and decision cycles accelerated, rather than a force judged mainly by its own organic kill count.

---

1. U.S. Marine Corps, *2025 Force Design Update* (Washington, DC: Headquarters Marine Corps, October 2025), https://www.marines.mil/Portals/1/Docs/Force_De sign_Update-October_2025.pdf.

CHAPTER 4
AVIATION AS AN IMPACT FORCE MULTIPLIER

Marine aviation is transitioning from a platform-centric enterprise to an impact-centric one, where value is measured less in sorties and tonnage and more in how aviation accelerates decisions, sustains distributed operations, and amplifies the lethality of the joint force.

The 2026 Marine Aviation Plan, anchored in Project Eagle, codifies this shift by treating aviation as an integrated enabler of stand-in forces and kill webs, not just an organic fires provider. It reframes the Aviation Combat Element (ACE) as a distributed, data-driven, sustainment-credible network that makes the entire Marine Air-Ground Task Force (MAGTF) faster, more survivable, and more lethal in contested environments.[*]

This chapter develops that logic in four steps. First, it explains how Project Eagle's three-horizon construct and lines of effort create strategic coherence for aviation as an impact multiplier across crisis response and the future fight. Second, it traces how Distributed Aviation Operations (DAO) and Decision-Centric Aviation Operations

---

[*] U.S. Marine Corps, Deputy Commandant for Aviation, *2026 Marine Corps Aviation Plan* (Washington, DC: Headquarters Marine Corps, February 10, 2026).

(DCAO) translate that strategy into operational practice, especially for stand-in forces in contested maritime spaces. Third, it examines the elevation of Aviation Ground Support (AGS) and the transformation of aviation sustainment as the decisive preconditions for real distributed impact. Finally, it reframes key platforms and capabilities, manned and unmanned, as impact nodes within joint kill webs rather than discrete, stove-piped commodities.

PROJECT EAGLE: ALIGNING AVIATION WITH IMPACT

Project Eagle provides the strategic blueprint that enables Marine aviation to function as an impact multiplier across time, rather than as a set of disconnected program decisions. It stretches planning across three Future Years Defense Programs (FYDPs): FYDP-1 (2026–2030) as "Fight Tonight," FYDP-2 (2031–2035) as "Bridge the Gap," and FYDP-3 (2036–2040) as "Future Fight."

Each horizon links specific modernization moves, readiness priorities, and capability transitions to a coherent theory of victory in contested environments. This temporal structure is not merely budget discipline; it is what allows aviation to deliberately sequence capability changes so that distributed, decision-centric operations grow more robust year over year.

Under Project Eagle, five lines of effort (LOEs) describe how aviation becomes an impact multiplier: concepts (DAO and DCAO), functions (including AGS as a seventh aviation function), transformation and innovation (especially AI/ML integration), resourcing across FYDPs, and platform/capability roadmaps.

Together, these LOEs ensure that doctrinal ideas, organizational functions, and investment decisions reinforce each other instead of diverging under separate bureaucratic logics. The result is an aviation enterprise designed to operate as a "team of teams," where fixed-wing, rotary-wing, tiltrotor, unmanned systems, and AGS converge on the problem of enabling stand-in forces to persist, sense, and strike under anti-access/area denial (A2/AD) pressure.

Project Eagle also explicitly links aviation's modernization to broader Service-level initiatives like Force Design and Project Dynamis. It positions aviation not as a niche community pursuing its own technology agenda, but as a principal contributor to the Corps' role in combined joint all-domain command and control (CJADC2) and naval expeditionary operations.

In that sense, the Aviation Plan is as much about cultural and organizational change, how Marines think about aviation's purpose, as it is about hardware, software, or structure. Aviation becomes the connective tissue that gives the MAGTF resilience and tempo inside an adversary's weapons engagement zone.

DISTRIBUTED AVIATION OPERATIONS: POSTURE FOR IMPACT

Distributed Aviation Operations are the central operational mechanism by which aviation becomes an impact multiplier for stand-in forces. DAO disperses aviation assets across hubs, spokes, and nodes to increase survivability, complicate enemy targeting, and provide persistent support to the Stand-In Force (SIF) operating inside contested maritime spaces. It is more than geographic dispersal; DAO demands that aviation sustain meaningful combat power from multiple expeditionary locations under contested logistics and electromagnetic attack.

In practice, DAO requires aviation to function as a networked set of expeditionary sites rather than a handful of large, predictable bases. Hubs provide semi-permanent infrastructure, full maintenance, robust command and control (C2), and deep fuel and munitions stocks.

Spokes offer more austere, but still sustainable, locations that can support flight operations for days or weeks. Nodes are intentionally transient and time-bounded, locations that exist for a mission or a narrow window of opportunity before they are displaced or shut down. This hub-spoke-node taxonomy is essential to deny the adversary lucrative, persistent targets while preserving aviation's ability to mass effects when required.

DAO's contribution to impact lies in how it underwrites the stand-in force concept. Stand-in units must survive, sense, and strike under persistent surveillance and long-range precision fire threat.

Aviation, operating through DAO, gives those units the ability to move across the battlespace, regenerate combat power, and maintain situational awareness without relying on a small number of vulnerable airfields.

It shifts aviation from "last-mile" support to a central determinant of whether stand-in forces can actually persist and remain lethal over time. In effect, DAO is aviation's operational answer to the question: how do we remain inside the adversary's weapons fan and still win.

DAO also changes how planners think about risk and time. Nodes, by definition, have shelf lives. They should not be treated as fixed positions to be defended indefinitely, but as temporal assets whose value is maximized when their timelines are deliberately planned and ruthlessly adhered to.

This timestamped view of nodes forces a more disciplined approach to movement, sustainment, and exposure, helping aviation conserve combat power for decisive windows rather than bleeding readiness in static, predictable postures.

DECISION-CENTRIC AVIATION OPERATIONS: DECISION AS THE WEAPON

If DAO answers "where" aviation operates, Decision-Centric Aviation Operations address "how" aviation makes decisions fast enough to matter across that distributed geometry. DCAO is the aviation-specific contribution to Project Dynamis and the wider CJADC2 effort: a shift from platform-centric planning to data-centric warfighting.

It rests on a simple premise: in peer conflict, the side that can sense, process, share, and act on information faster wins. The ability to close kill webs repeatedly and reliably is at least as decisive as the raw number of aircraft or munitions.

DCAO aims to turn aviation into a data-rich, AI-enabled decision enterprise. That means treating every platform—not only fifth-generation fighters and long-endurance UAS, but also tiltrotors, heavy-lift helicopters, and rotary-wing aircraft, as a sensor and data relay node. It means fusing maintenance, supply, operational, and C2 data into architectures where AI/ML tools support commanders and staff with forecasting, pattern recognition, and course-of-action development. And it means building digital interoperability into aviation systems so that data moves with low friction across Services, platforms, and domains.

The cultural implications are significant. Modern C5ISR systems give senior commanders unprecedented visibility into tactical operations, creating a temptation to centralize decision-making. But in a contested electromagnetic spectrum, high-tempo fight, such centralization can become a liability as bandwidth collapses and human decision cycles lag behind machine-speed engagements.

DCAO therefore pushes decision authority downward, pairing better situational awareness at the edge with mission command principles so tactical formations can act within the commander's intent without waiting on remote approval for every move.

In this context, aviation multiplies impact when it shortens the sensor-to-decision-to-shooter cycle for the entire joint force.

- Fifth-generation fighters that discover targets and cue joint fires without expending their own weapons have multiplied impact if their data enables more lethal, precise, and timely fires from naval, ground, or allied platforms.
- Long-endurance UAS that maintain pattern-of-life awareness over key maritime corridors enable commanders to shape the battle days in advance.
- Rotary-wing aircraft that serve as airborne gateways, translating among disparate data links and voice nets, allow legacy systems to participate in kill webs that would otherwise exclude them.

DCAO is thus less about a set of discrete "digital" projects and more about a mindset: aviation's primary weapon is the networked information it generates and moves, with kinetic fires as one of several ways to translate decision advantage into effects.

AVIATION GROUND SUPPORT AND SUSTAINMENT: THE HIDDEN CENTER OF GRAVITY

The 2026 plan's decision to elevate Aviation Ground Support to the seventh function of Marine aviation reveals a deep appreciation of what impact forces actually require. Distributed, decision-centric operations are unsustainable if the ACE cannot generate fuel, power, maintenance, mobility, and expeditionary infrastructure at the pace and scale DAO demands. AGS is the backbone that turns DAO from a PowerPoint concept into an operational reality.

Treating AGS as a core aviation function rather than a supporting afterthought forces the enterprise to invest in its recapitalization. That includes modernizing expeditionary fuel and power systems, airfield services, mobility platforms, and protection measures for distributed sites.

It also means ensuring AGS units are fully integrated into planning cycles for DAO and DCAO, rather than being told late in the process to "enable" aviation decisions already made elsewhere. In a truly distributed fight, the feasibility of a DAO concept is determined first by AGS capacity, not by the number of aircraft available on paper.

Parallel to AGS's elevation is a broader transformation of aviation sustainment. The plan is explicit that Marine aviation can no longer afford reactive maintenance, stovepiped supply chains, and labor-intensive, manual planning for flight operations.

To sustain distributed aviation in contested environments, the ACE must become a predictive, data-enabled sustainment enterprise. That transformation runs along three main lines of operation: dynamic aviation supply, predictive maintenance, and optimized operations.

Dynamic aviation supply rethinks supply packages and support concepts for a nodal, fluid architecture instead of a static base structure. It leverages AI/ML to analyze configuration and failure data, especially for high-end systems like the F-35, and to refine afloat and base spare packages for deployments.

The goal is higher readiness, reduced material footprint, and better alignment between supply posture and DAO patterns. Predictive maintenance applies AI/ML tools to maintenance data across multiple Type/Model/Series, enabling maintainers to anticipate failures, reduce man-hours on unnecessary inspections, and increase aircraft availability.

Optimized operations fuses data from historically separate systems, maintenance, readiness, and logistics information systems, into AI-enabled tools that support scheduling and operational planning. By automating complex, data-heavy tasks, these tools free commanders and staff to focus on warfighting problems rather than spreadsheet management. In aggregate, these sustainment initiatives seek to create an aviation enterprise that can sustain high-tempo distributed operations without being consumed by its own support requirements.

Importantly, the sustainment shift includes a heavy training and manpower component. Advanced Maintenance Training Academies, updated courses for maintenance chiefs, and tighter alignment with training commands and industry partners are all directed toward building a workforce comfortable with data-rich tools and advanced systems.

Manpower initiatives focus on aligning talent with modernization pathways, preserving expertise in legacy platforms through transition, and retaining critical skills with appropriate incentives. Aviation impact depends as much on the human system as on any AI algorithm or new platform.

PLATFORMS AND CAPABILITIES AS IMPACT NODES

Viewed through an impact-force lens, platform discussions shift from "what can this aircraft do" to "how does this aircraft contribute to joint kill webs and distributed operations." The 2026 plan's commodity sections, fixed-wing, tiltrotor, rotary-wing, unmanned systems, support aircraft, weapons, digital interoperability, and survivability, can thus be read as a catalog of impact nodes rather than an inventory list.

Fifth-generation fighters exemplify this shift. Their value is no longer captured solely by the number of weapons released or targets struck in a sortie. Instead, their real contribution emerges when they act as highly survivable sensor-shooter-nodes: discovering targets beyond the reach of other sensors, fusing multi-domain data into a coherent picture, managing portions of the kill web as airborne C2, and selectively striking when timelines or weapon effects demand it. A mission where a fighter discovers and shares multiple high-value targets for joint fires without firing a shot can be more consequential, in impact terms, than a mission where it independently services a smaller target set.

Long-endurance unmanned aircraft, such as medium-altitude, long-endurance (MALE) systems, offer persistent ISR and communications relay that underpin DCAO across large maritime theaters. Operating from expeditionary sites, they provide the MAGTF and joint force with the ability to maintain continuous sensing, validate patterns of life, and keep kill webs supplied with fresh information. As the aviation roadmap progresses toward more advanced UAS and family-of-systems constructs, these platforms become central to the Corps' ability to hold adversary assets at risk across time and space.

Tiltrotor and heavy-lift assault support platforms, such as MV-22 and CH-53K, are impact enablers through movement and sustainment. Their ability to move Marines, equipment, and sustainment loads across wide, contested areas allows commanders to establish, reinforce, and displace hubs, spokes, and nodes at the tempo DAO requires. Heavier lift to austere locations, combined with improved survivability and digital integration, changes the logistics calculus for distributed forces and expands the menu of feasible operational designs.

Legacy rotary-wing platforms gain renewed relevance when they are fully integrated into the digital ecosystem. With robust digital interoperability, H-1 aircraft can serve as airborne gateways, close air support platforms, and sensor relays within kill webs that also include fifth-generation fighters, UAS, and naval fires. The plan's commitment to sustain and modernize these platforms—even as future attack and strike capabilities emerge—reflects a recognition that impact comes from integration as much as from raw performance metrics.

Weapons, spectrum, and survivability upgrades round out the impact picture. Advanced long-range precision munitions, improved aircraft survivability equipment, and digital interoperability systems all contribute to aviation's ability to generate effects from dispersed positions while remaining survivable under sophisticated threat. These investments ensure that aviation nodes within the kill web are not only connected, but also capable of delivering meaningful, credible fires when required, even as they prioritize sensing and decision-support roles.

FROM AVIATION TO IMPACT FORCE

The 2026 Marine Aviation Plan, viewed through this lens, is less about "fixing aviation" and more about transforming the ACE into an impact force multiplier embedded in joint kill webs.

Project Eagle's three-horizon framework prevents modernization from becoming a series of disconnected bets by anchoring it to a coherent theory of distributed, decision-centric operations. DAO provides the posture. DCAO provides the decision advantage. AGS and predictive sustainment provide the foundation without which distribution collapses under its own weight. Platforms and capabilities, manned, unmanned, and enabling, become nodes in a larger system of systems whose purpose is to give the Marine Corps and the joint force a persistent, survivable, and agile presence inside contested spaces.

For defense professionals evaluating this trajectory, the central question is no longer whether Marine aviation can "support the ground scheme of maneuver" in the traditional sense.

The more relevant question is how aviation, configured under Project Eagle, can shape the battlespace, compress decision cycles, and enable stand-in forces and naval partners to fight and win inside the arc of adversary weapons and sensors.

An aviation enterprise that answers that question convincingly will have moved beyond platform metrics to genuine impact, and in doing so, will sit at the heart of the Corps' contribution to future conflict.

CHAPTER 5

KILL WEB IMPACT: C2, DIGITAL INTEROPERABILITY, AND AIR DEFENSE

The impact-force concept rests on kill webs, networks in which many sensors, decision nodes, and shooters connect dynamically so that any vetted sensor can cue any appropriate shooter through multiple pathways.

This differs fundamentally from traditional kill chains, where a linear find–fix–track–target–engage–assess sequence collapses when any single link breaks.

For small, forward Marine detachments with limited organic firepower, kill webs are what turn presence into impact they allow those units to discover and characterize targets, feed information through resilient networks, and cue naval, air, cyber, or allied fires they do not own. Without kill-web integration, the same forces are reduced to the effects they can generate with the weapons on their shoulders or on their vehicles.

Steel Knight 25 was one of the first major exercises to test whether Marines could actually operate within kill webs at something approaching campaign tempo. What emerged was an institution that understands the kill-web problem intellectually and has begun to build the building blocks, Distributed Aviation Operations, Decision-Centric Aviation Operations, digital interoperability, MAOC, MADIS, MRIC, but still wrestles with command-and-control habits, cultural reflexes, and sustainment realities inherited from the crisis-response era.

FROM CRISIS KILL CHAINS TO CHAOS KILL WEBS

Kill webs are not simply a more complicated way to describe targeting. They are the operational grammar of a force designed for chaos management rather than episodic crisis response.

Crisis-era forces could surge from sanctuary, concentrate mass against a discrete problem, and then stand down once the situation stabilized. In that environment, linear kill chains that tied sensors to shooters within coherent task forces were adequate. The emerging environment is very different overlapping gray-zone pressure, contested logistics, degraded space and cyber architectures, and political decision cycles shaped in real time by information operations.

In that world, the impact force must live forward, persist under friction, and generate effects even when higher headquarters are degraded and communications are intermittent.

Kill webs provide the connective tissue for that kind of force. They allow a small stand-in unit on an island or littoral strip to act as the joint force's forward eyes, ears, and brain stem rather than as an isolated missile battery. The unit's organic weapons still matter, but its real value lies in its ability to discover, classify, and frame problems faster than the adversary, then enable joint and allied fires to exploit that decision advantage.

Steel Knight 25's campaign design, embassy reinforcement under pressure, noncombatant evacuation, distributed fires, contested logistics, and kill-web C2 across hundreds of miles, was written explicitly to test whether the Marine Corps could move from talking about that logic to living it.

G-6 "THREE WAYS TO WIN" AND THE DIGITAL BACKBONE

Within 3rd Marine Aircraft Wing, the G-6 staff translated abstract kill-web diagrams into a practical communications architecture built on a simple rule of thumb: "I need three ways to win, three ways of getting a message across the wire." Their task was to separate sensors from shooters, because weapons now outrange the sensors mounted on individual platforms, while preserving the speed and precision necessary for contested combat.

The solution was not a single exquisite pipe, but an ecosystem of pathways: Link-16 for tactical data exchange, military SATCOM for protected beyond-line-of-sight connectivity, commercial SATCOM such as Starlink or ViaSat for additional capacity, cellular and fiber where available, and unclassified networks to move nonsensitive sensor data rapidly.[*]

The 2026 Marine Aviation Plan codifies that logic in its description of Digital Interoperability and the MAGTF Agile Network Gateway Link. It defines four pillars, sensors, processors, interfaces, and radios, and treats DI/MANGL as a family of systems that can integrate disparate platforms and waveforms into a common architecture rather than a monolithic solution. Extended Tactical Network concepts and MAGTABs expand that architecture to dismounted Marines, vehicles, and command posts, enabling them to share a common picture up to the Secret level and to use unclassified networks where speed and agility matter more than high classification.

Yet Steel Knight made clear that this digital backbone creates a paradox. Networked forces can make decisions faster, but every transmission generates electromagnetic exposure that sophisticated adversaries can detect and target. One participant noted that a 32-second radio transmission could bring adversary fires within two minutes, compressing decision cycles beyond what hierarchical command structures can comfortably manage. The kill web, in other words, is both an

---

[*] See the interview in part two, chapter seven in this book.

enabler and a signature. Managing that tension is now a core task for the impact force.

DAO, DCAO, AND AVIATION
AS KILL-WEB ENGINE

Project Eagle, the framework introduced in the 2026 Aviation Plan, treats Distributed Aviation Operations and Decision-Centric Aviation Operations as the aviation engine of the impact force. DAO is defined as more than scattering aircraft; it is the deliberate design of an aviation network built on expeditionary hubs, spokes, and forward arming and refueling points that can sustain combat power under adversary targeting. It elevates Aviation Ground Support to the seventh function of aviation and sets specific modernization lines for expeditionary arresting gear, transportable firefighting and refueling systems, and C-130-movable infrastructure.

Steel Knight 25 showed DAO moving off the slide. 3rd MAW ran aviation spokes from main bases to places like Sacramento Mather and Victorville, where Marine Wing Support Squadrons turned civilian airfields into functioning nodes capable of hot refuel, re-arm, maintenance, and basic C2. MAG-16's MV-22s and CH-53s used those nodes to execute long-range insertions and logistics, while MAG-11's F-35Cs treated their presence at an expeditionary strip less as a novelty and more as a working model of how fifth-generation aircraft will operate under DAO.

DCAO is the cognitive counterpart. The plan links it to Project Dynamis and Combined Joint All-Domain Command and Control, framing it as the aviation-specific effort to use data and AI/ML decision support to increase the speed and quality of decisions across distributed forces. In practice at Steel Knight, DCAO meant F-35s, MQ-9s, MV-22s, CH-53s, and H-1s all acting as sensors, shooters, and C2 nodes in kill webs rather than as self-contained platforms. It also meant beginning to fuse maintenance, supply, and operations data through AI-enabled tools so that aviation commanders could anticipate readiness, schedule more intelligently, and sustain DAO under stress.

The lesson was straightforward: distribution only increases impact if the distributed force can think faster than the adversary. Kill webs are what allow DAO and DCAO to converge, turning a dispersed set of aircraft and detachments into a coherent, decision-advantaged aviation combat element.

MAOC, MACG-38, AND GROUND-BASED AIR DEFENSE

Command and control is where kill-web theory collides most directly with institutional habit. The 2026 plan pushes toward Multifunction Air Operations Centers that consolidate the functions of the old Direct Air Support Center and Tactical Air Operations Center into a unified capability optimized for distributed employment. By 2029, active and reserve MACGs will be reorganized around MAOC and modernized Marine Air Control Squadrons, with CAC2S, small-form-factor systems, and updated air-traffic-control capabilities forming an integrated MACCS.

Colonel O'Connell's MACG-38 at Steel Knight provided a preview. Instead of treating his group as a back-office enabler, he framed it as the dial that configures aviation for distributed operations—combining MWSS, MASS-3, MACS-1, LAAD, and comms elements into modular AC2GS packages for specific missions. Those packages created what the impact-force logic requires: small, mobile command posts with enough sensors, communications, and air-defense capability to make decisions, protect themselves, and feed the wider kill web without becoming stationary, high-signature magnets for long-range fires.[*]

Ground-based air defense is both shield and node in this system. The Marine Air Defense Integrated System, its lighter L-MADIS variant, and the Medium Range Intercept Capability together create a layered defense against UAS, cruise missiles, and manned aircraft, anchored by TPS-80 G/ATOR and the emerging Medium Range Air Defense Radar. The Aviation Plan lays out a growth path to 190

---

[*] See the interview in part two, chapter eight in this book.

MADIS systems, 21 L-MADIS, and three MRIC batteries by the early 2030s.

Steel Knight revealed the tension this creates. GBAD units that transmitted continuously contributed richly to the kill web but lit themselves up electromagnetically; when they went emissions-controlled to preserve survivability, they risked becoming blind and isolated.

The impact force will not escape this paradox entirely, but it can manage it better by using MAOC-led C2 packages to choreograph emission control, exploit unclassified and low-probability-of-intercept links, and accept that air defense nodes will sometimes operate primarily as protectors and sometimes as contributors to broader kill webs.

ROTARY-WING AND
UAS AS DIGITAL NODES

Rotary-wing and unmanned systems are where kill webs become tactically tangible. The 2026 Aviation Plan's H-1 modernization path, SPINE for power, Precision Attack Strike Munition, AIM-9X, and DAIRCM for survivability, with Link-16 and proliferated LEO connectivity for digital interoperability, is written against a vision of Cobras and Hueys as low-altitude kill-web nodes.

HMLA-267 at Steel Knight was that vision in its first operational form: a fully DI squadron using Link-16 and MAGTABs to pull F-35 and MQ-9 sensor data, share its own targeting inputs, and act as a C2 extension for stand-in ground forces.[*]

Major Jonathan Moss's journey from traditional voice-radio Cobra pilot to digital-interoperability lead illustrates the human adaptation that underpins this shift. At VMX-1 he helped test how H-1s could use data links to compress the targeting cycle, integrate with fifth-generation aircraft, and manage small unmanned systems; at HMLA-267 he is now turning that experience into tactics, training syllabi, and culture for an operational squadron. The squadron's ability to operate from

---

[*] See the interview in chapter five in part two of this book.

austere sites, plug into kill webs, and still provide immediate close support shows why rotary-wing aviation remains central to chaos management once it gains the ability to sense, share, and survive in contested airspace.[*]

Unmanned aviation extends the web's reach. The Aviation Plan describes MQ-9A as a strategic sensor and kill-web enabler across the Pacific, with Increment II payloads—Sky Tower airborne relay, electronic-support pods, pLEO SATCOM, maritime-domain-awareness sensors, and detect-and-avoid systems, designed to turn VMUs into persistent C2 and sensing hubs. Steel Knight's experiments with integrating unmanned surface vessels into the air picture via unclassified networks, and proliferating their data so any shooter could exploit it, were early glimpses of this logic at work. Over time, the convergence of MQ-9, MUX family systems, and MAGTAB-enabled ground nodes will make it increasingly normal for an autonomous sensor to cue a manned shooter or vice versa through multiple, redundant pathways.

CULTURE, AUTHORITIES, AND THE HUB-VERSUS-NODE PROBLEM

Beneath the technology runs a more stubborn issue: whether the Marine Corps is willing to operate as a truly nodal force or simply as a set of disaggregated hubs. Steel Knight's physical distribution was impressive; units and nodes stretched from the coast to inland spokes and island ranges. But decision authority and information flow often remained hub-centric, with MEF and wing headquarters retaining control as long as communications held and subordinate elements waiting for guidance when links degraded.

Lieutenant General Hedelund's notion of timestamped nodes, positions conceived with inherent shelf lives rather than as quasi-permanent bases, offered one solution, encouraging commanders to think of nodes as temporary, effect-focused constructs rather than static assets. The Aviation Plan's emphasis on S^2D or survivability, sustainability, duration as the framework for employment of aviation nodes echoes

---

[*] See the interview in chapter six in part two of this book.

that logic. But neither fully resolves the deeper tension between an institutional preference for centralized, risk-managed control and the operational need for decentralized initiative when a 32-second transmission can get a node targeted.*

The Ukrainian experience, frequently cited during Steel Knight, underscores what is at stake. Ukrainian units have demonstrated the ability to integrate commercial drones, adapt tactics, and invent new combined-arms approaches faster than their adversaries, often without waiting for doctrinal approval. That speed of adaptation, enabled by trust in junior leaders and comfort with improvisation, is precisely what kill webs demand: if every sensor can feed every shooter, hierarchical command becomes a bottleneck rather than an advantage.

For the Marine Corps, the kill-web problem is therefore as much about authorities and culture as it is about networks and platforms. The 2026 Aviation Plan reflects that recognition by anchoring its modernization blueprint in the T-E-A-M philosophy, elevating human factors, safety, and disciplined execution of basics as prerequisites for operating complex digital architectures under stress. Steel Knight 25, in turn, revealed how much work remains to align mission-type orders, risk tolerance, and joint targeting authorities with a world in which a sergeant at a forward node may see and understand the fight before higher echelons do.

The impact force will only be as decisive as its kill webs are real. Technically, Project Eagle, DAO/DCAO, DI/MANGL, MAOC, MADIS, MRIC, and H-1/MQ-9 modernization are building a credible foundation.

Operationally, exercises like Steel Knight are beginning to expose and close gaps in C2, sustainment, and EM-signature management. The remaining question is whether the Corps can build the trust, establish the authorities, and internalize the cultural changes necessary to let distributed Marines use those kill webs at the speed that chaos management demands.

---

* See the interview in chapter two in part two of this book.

CHAPTER 6

PLATFORMS IN AN IMPACT FORCE: FROM INVENTORY LISTS TO CAMPAIGN EFFECTS

The 2026 Marine Aviation Plan is, on the surface, a platform-centric document: pages of F-35B/C inventory targets, MQ-9 squadron growth, MV-22 and CH-53K timelines, H-1 upgrade paths, and weapon integration schedules.

Yet that is not how Marines at Steel Knight 25 thought about aviation. When viewed through an impact force lens, the central question shifts from "What can this aircraft do?" to "How does this platform generate disproportionate effects within joint kill webs and a chaos-management campaign?"

That distinction is not rhetorical. An F-35 sortie that discovers ten targets, distributes that data across a joint kill web, and enables dozens of successful strikes by other shooters has greater operational impact than an F-35 that drops its own weapons on two targets and then goes home. Steel Knight 25 made that difference visible in real time, as aircraft and command-and-control nodes were tasked less as independent shooters and more as sensing, fusing, and enabling nodes within a distributed campaign.

This chapter walks through four key aviation platforms, F-35, MQ-9, MV-22/CH-53K, and H-1, as case studies in impact. Each section anchors the 2026 Aviation Plan's programmatic logic in what

Marines actually tried to do at Steel Knight 25 and how those experiences are reshaping both expectations and employment concepts.

F-35: FROM SHOOTER TO SENSOR-SHOOTER-NODE

By the end of 2026 the Marine Corps plans to field roughly 205 F-35Bs and 56 F-35Cs, organized into 14 fleet squadrons and supported by test and fleet replacement units. In the programmatic language of the Aviation Plan, this is about transition timelines, procurement quantities, and readiness rates. In the field, however, the F-35 is increasingly treated less as a boutique strike fighter and more as an airborne brain for the air-ground team.

At Steel Knight 25, MAG-11's F-35Cs were employed as multi-role nodes: advanced sensors, airborne C2 extensions, electronic-warfare platforms, data-fusion engines, and, when necessary, time-sensitive strike assets. F-35 crews were as focused on what their sensor picture could do for the rest of the force as on the weapons they personally carried.

In several vignettes, the measure of success was how many lethal firing opportunities the F-35s could create for other shooters, Navy surface combatants, joint air platforms, and ground-based fires, rather than their own bomb damage assessments.

Seen from an impact-force perspective, the F-35's core contributions fall into several categories:

- Penetrating sensor: operating inside threat rings to locate and characterize targets whose detection exceeds the survivability envelope of legacy platforms.
- Airborne C2 node: managing distributed aviation packages, de-conflicting fires, and acting as a forward extension of wing and MAGTF command-and-control architectures.
- Kill-web catalyst: passing high-quality targeting data to Navy ships, Air Force bombers, and ground units, turning a single F-35 sortie into a multiplicative force for joint fires.

- Electronic-warfare contributor: degrading adversary sensors and networks while preserving the integrity of Marine and joint connectivity.
- Organic shooter of last resort: employing its own weapons when timelines or communications gaps make external cueing impractical.

The 2026 plan's Project Eagle framework, Fight Tonight (2026–2030), Bridge the Gap (2031–2035), and Future Fight (2036–2040), codifies this transition by emphasizing digital interoperability, decision-centric operations, and distributed aviation as core F-35 employment concepts rather than add-ons. Steel Knight 25 validated both the promise and the friction: F-35s could dramatically extend the reach and tempo of the kill web, but they also exposed shortfalls in gateway systems, data standards, and TTPs for managing information flow without overwhelming operators.

In practical terms, the F-35 has become the exemplar of what an impact-force platform looks like: a system whose greatest value lies not in the number of weapons it can carry, but in the speed and quality of decisions it enables across the joint force.

MQ-9: PERSISTENT EYES AND THE FABRIC OF DECISION ADVANTAGE

If the F-35 provides exquisite, penetrating sensing and fusion, the MQ-9 brings the kind of persistence that manned platforms simply cannot match. The 2026 Aviation Plan expands Marine employment of MQ-9s, with VMU squadrons operating from expeditionary locations to provide continuous intelligence, surveillance, and reconnaissance across wide areas.

An MQ-9 orbit does not, by itself, produce kinetic effects. Its impact lies in creating a continuous, high-fidelity understanding of patterns of life, movement corridors, and emerging threats. With endurance often exceeding 20 hours, a single MQ-9 can maintain custody of targets and areas of interest through multiple decision cycles, allowing commanders to shift from episodic "snapshot" awareness to something closer to a living, updated picture.

During Steel Knight 25, MQ-9 feeds were integral to the campaign-laboratory design. Embassy reinforcement and noncombatant evacuation scenarios, distributed fires problems, and contested logistics vignettes all depended on persistent sensing to identify threats to routes, staging bases, and forward nodes. In several cases, the value of MQ-9 was less about feeding a single firing solution and more about enabling commanders to shape the timing, sequencing, and direction of operations based on evolving adversary behavior.

From an impact-force standpoint, MQ-9s contribute in at least three ways:

- Continuous ISR: generating unbroken coverage over key areas, filling in the gaps between manned sorties and satellite revisits.
- Target development: building targetable patterns that joint fires can exploit efficiently, reducing the number of sorties and munitions required to achieve desired effects.
- Network backbone: acting as communications relays and gateways to extend kill-web connectivity across dispersed nodes and austere locations.

The Aviation Plan reflects this by prioritizing MQ-9 configurations that emphasize sensors, communications payloads, and integration with Marine kill-web architectures rather than treating the platform primarily as a "strike drone." Steel Knight 25 underscored the logic: in a chaos-management environment, persistent awareness is itself a form of combat power, because it allows Marines and joint commanders to decide faster, allocate scarce assets better, and avoid being surprised.

MV-22 AND CH-53K: MOBILITY AS A WEAPON IN DISTRIBUTED AVIATION OPERATIONS

Assault support platforms, the MV-22 Osprey and CH-53K King Stallion, do not appear on traditional lists of "lethal" systems. Their impact derives from movement: inserting, sustaining, repositioning, and extracting forces across contested geography. In a distributed aviation construct, mobility is as decisive as firepower.

Over two decades, the MV-22 has evolved from a controversial replacement for legacy helicopters into the backbone of Marine operational and strategic reach. Marines used the Osprey first as a high-speed assault platform, then as a crisis-response connector, and now as a key enabler of distributed operations across theater-scale distances.

At Steel Knight 25, MV-22s were central to linking expeditionary airfields, forward arming and refueling points, and dispersed ground nodes across the American Southwest, demonstrating how speed and range turn geography from a constraint into an opportunity.

The CH-53K introduces a different kind of impact multiplier: sheer lift capacity. One CH-53K sortie can deliver heavy payloads, vehicles, bulk fuel, ammunition, power modules—to austere locations that would require multiple MV-22 sorties or ground convoys under other circumstances. In a distributed aviation operations (DAO) framework, this matters enormously. A single heavy-lift mission can establish a functioning node, a FARP, a sensor site, a forward command post, that then enables weeks of operations for other platforms.

Steel Knight 25, however, also illuminated the gap between aspiration and reality. The exercise relied largely on CH-53Es, with only limited CH-53K availability, underscoring how much current DAO logistics planning still depends on legacy capacity and how transformational full CH-53K fielding could be.[*] Marines repeatedly high-

---

[*] CH-53K King Stallions were employed during Steel Knight 25, including in heavy-lift roles such as lifting a Joint Light Tactical Vehicle (JLTV) as part of the exercise conducted across West Coast locations in early December 2025.

lighted how AO design and mission profiles would change once the King Stallion's lift and range are fully available, particularly for sustaining distributed nodes deep inside contested areas.

In a chaos-management context, MV-22s and CH-53Ks generate impact in several ways:

- Survivable insertion: delivering forces into dispersed, non-traditional landing zones, complicating adversary targeting and reducing dependence on fixed infrastructure.
- S^2D logic (survivability, sustainability, duration): shaping operations around how long nodes can be maintained under contested logistics, not just whether they can be initially inserted.
- Dynamic repositioning: enabling commanders to shift weight, reconstitute forces, and exploit opportunities faster than adversaries can adjust.

The 2026 Aviation Plan elevates Aviation Ground Support and logistics as core functions within DAO, recognizing that platforms like MV-22 and CH-53K only create impact if fuel, munitions, parts, and maintenance can move with similar agility. Steel Knight 25's contested logistics vignettes made this explicit, forcing planners to confront the real S^2D calculus of distributed operations rather than assuming benign sustainment conditions.

H-1: LEGACY AIRFRAMES TURNED DIGITAL GATEWAYS

On paper, the H-1 family, AH-1Z Viper and UH-1Y Venom, looks like a legacy rotary-wing solution held over until a "future attack" platform arrives. In practice, the H-1 community has become one of the clearest illustrations of how digital integration can turn an existing aircraft into an impact-force node.

HMLA-267's journey is emblematic. Under leaders who had lived both the analog Cobra world and the F-35's digital environment, the squadron drove a full-spectrum digital interoperability upgrade:

common waveforms, data-link integration, networking gateways, and procedures that allowed H-1s to plug into the same kill webs as fifth-generation fighters and unmanned systems. The result is a very different platform profile:

- Digital gateway: bridging ground units, legacy radios, and advanced networks, ensuring that even small detachments can access and contribute to the broader kill web.
- ISR contributor: using sensors and data links to provide real-time targeting and situational awareness at the edge, rather than relying solely on voice reports.
- Communications relay: extending connectivity into dead ground or complex terrain where line-of-sight links are unreliable.
- Traditional fires: still delivering close air support and armed escort, but now with better targeting data, de-confliction, and coordination.

The 2026 Aviation Plan reflects this shift by committing to keep H-1 viable into the 2040s with lethality, survivability, and digital-interoperability upgrades, even as future attack/strike studies proceed. In other words, H-1s are being modernized not to compete with the F-35 or MQ-9 on raw performance, but to enhance their impact as connective tissue in the kill web.

Steel Knight 25 underscored how valuable this connective role can be. In scenarios where ground units were dispersed, communications were degraded, and higher-echelon command posts were deliberately austere to reduce their electromagnetic signatures, H-1s often provided the practical link between what Marines on the ground were seeing and what the wing or joint fires enterprise could act upon. The platform had not changed; the way it was integrated had.

INTEGRATING THE CASE STUDIES: PLATFORMS AS NODES IN A KILL WEB

Viewed individually, each platform presents a familiar narrative of modernization: more F-35s, expanded MQ-9 employment, CH-53K fielding, H-1 upgrades. Viewed together, in light of the 2026 Aviation Plan and Steel Knight 25, a different pattern emerges. The platforms gain their real meaning as nodes in a kill web and as tools for chaos management rather than as isolated capabilities.

Steel Knight 25 made this concrete. F-35s discovered and character-ized targets at range. MQ-9s maintained custody and fed continuous updates as the situation evolved. H-1s connected dispersed ground units to those targeting pictures and provided immediate fires when required. MV-22s and CH-53Es (and, in future, CH-53Ks) moved the people, fuel, munitions, and power that made those nodes real in unforgiving terrain. Marine Air Control Group-38 and emerging command-and-control constructs at the Wing Operations Command Center stitched these pieces together under conditions of deliberate friction, jamming, bandwidth limitations, and incomplete tasking.

The Aviation Plan's emphasis on Distributed Aviation Operations and Decision-Centric Aviation Operations provides the doctrinal scaf-folding for this integration: distribution not for its own sake, but to preserve combat power under attack, and decision-centricity to ensure that distribution actually generates faster, better choices rather than confusion. AI/ML sustainment initiatives, dynamic supply, predictive maintenance, optimized operations, aim to keep this distributed avia-tion enterprise flying in the face of contested logistics and high opera-tional tempos.

At the same time, both the plan and the exercise exposed the remaining obstacles:

- Interoperability gaps: Not all platforms can talk to all others;. Data formats and waveforms remain uneven across the fleet.
- Training emphasis: Squadrons are still incentivized to perfect

platform-specific skills more than cross-platform, networked employment.

- Logistics density: Steel Knight's contested logistics scenarios highlighted how thinly stretched sustainment assets become once distribution is taken seriously.

These frictions are not signs that the impact-force concept is flawed. They are the inevitable by-products of moving from PowerPoint to practice. The combination of an increasingly demanding Aviation Plan and exercises like Steel Knight 25 is forcing the Marine Corps to discover, platform by platform, what it actually takes to turn a crisis-response air arm into the aviation backbone of a stand-in, chaos-management impact force.

The case studies in this chapter suggest a common conclusion. The decisive question for Marine aviation is no longer, "How lethal is this platform on its own?" It is, "How much faster, more precise, and more resilient does this platform make the entire joint and allied force when it is treated as a node in a kill web, sustained by distributed logistics, and employed within an impact-force concept?"

CHAPTER 7

COMMANDING IMPACT: CULTURE, MISSION COMMAND, AND HUMAN FACTORS

Lieutenant General (Retired) Robert Hedelund framed the central problem for impact forces in a single question: how do we fight with the nodes we actually have, at the moment we actually have them?*

That question goes beyond platforms and sensors and cuts directly into culture, command philosophy, and the human factors that either unlock or throttle combat potential in a distributed fight. Technology can connect Marines across vast distances, but only the right habits of command and the right patterns of trust will allow them to exploit those connections under fire.

MISSION COMMAND UNDER PRESSURE

The Marine Corps has long embraced mission command in concept, but exercises like Steel Knight expose how hard it is to live that concept in practice.

Modern C5ISR architectures give senior commanders a level of real-time visibility that previous generations could not imagine, and

---

* See the interview in chapter two in part two of this book.

that visibility brings a constant temptation to direct instead of command.

Marines increasingly understand what distributed operations must look like in the Pacific, but they still wrestle with the how of letting go of centralized control. When a 32-second transmission can trigger hostile retribution within two minutes, the traditional instinct to centralize control becomes not just sub-optimal but dangerous.

Mission command in this environment requires a conscious act of restraint by higher headquarters. Leaders must deliberately choose to set clear intent, frame boundaries, and then stay out of the way long enough for subordinate commanders to maneuver inside that intent.

The more the common operational picture improves, the stronger the gravitational pull toward meddling in tactical decisions that should remain at lower echelons. Impact forces will succeed only if senior leaders can discipline themselves to use information to empower, not to hover.

THE DELEGATION DILEMMA

Distributed operations raise the stakes on delegation to a level that would have been unthinkable when firepower and sensing were more centralized. Today, relatively small units can command situational awareness and strike options that used to reside only in much larger formations, which turns every platoon and every detachment into a potential strategic actor.

Yet capability without corresponding authority and clear tasking becomes a form of friction, as Marines watch opportunities appear in their sensors that they are not empowered to seize.

For impact forces, the delegation dilemma is not abstract. Subordinate commanders will face moments in which they must make consequential decisions without waiting for confirmation, recognizing that feedback on whether they were "right" may not arrive until hours or days later.

That reality demands a culture that accepts reasonable risk, tolerates honest mistakes made within commander's intent, and resists the urge to rewrite every local decision from higher headquarters after the fact. When the system punishes initiative more harshly than it punishes inertia, Marines will correctly infer that the safer course is to wait rather than act.

Delegation in impact forces also depends on how clearly higher headquarters expresses priorities. If subordinate units understand which effects matter most, disrupting a kill chain, preserving particular nodes, creating specific dilemmas for the adversary, they can make decisions that align with campaign logic even when they cannot reach back for guidance. The more ambiguous the guidance, the more likely that distributed actions will fragment into isolated episodes rather than a coherent operational design.

CONTINUITY IN A CHANGING FORCE

Despite the eye-watering speed and connectivity of modern systems, the essence of Marine combined arms has not disappeared. Hedelund's reflection that much of what he saw in Steel Knight echoed what he did in 1985 captures an important continuity that should anchor cultural adaptation. The tools have changed, and decisions can be made faster, but the combined arms force remains the secret sauce: a command element, capable aviation, ground combat forces, and logistics, integrated into a single fighting system.

This continuity matters because Marines are not reinventing war from scratch in order to field impact forces. They are adapting a proven combined arms logic to an environment where digital interoperability, kill web architectures, and dispersed posture redefine how those elements connect.

Whether organized as a traditional MAGTF, as a Marine Littoral Unit, or under some future construct, the fundamental requirement remains a force that can sense, shoot, maneuver, and sustain as a coherent whole. Recognizing that continuity can prevent the force from chasing fads and help leaders focus on the real problem: how to marry enduring combined arms practice with new ways of distributing nodes and authorities.

COMMAND RELATIONSHIPS
AND CAMPAIGN LOGIC

If culture shapes what commanders are willing to delegate, command relationships shape what those delegated actions add up to. How do Marines plug into joint operations as both force providers and operational headquarters?

Steel Knight exposed the tensions inherent in 3rd MAW trying to serve simultaneously as a source of aviation capability and as a headquarters orchestrating those capabilities across a distributed battlespace.

Impact forces will generate tactical effects across the theater, but those effects only matter if they aggregate toward strategic objectives. Without clear command relationships, well-intended actions by dispersed units can easily become isolated events that consume scarce resources without shaping the campaign in meaningful ways.

Commanders must therefore design relationships and authorities so that every local engagement contributes to a larger operational narrative. That means clarifying who has the lead for integrating Marine aviation into joint fires, how Marine commanders present forces to the joint fight, and where the authority lies to shift those forces as the situation changes.

The challenge is not only structural but cognitive. Leaders have to think in terms of kill webs and networked effects rather than linear tasking and rigid apportionment. When they do, they can treat each distributed Marine node as a contributor to a shared operational design rather than as a self-contained problem to be deconflicted.

HUMAN FACTORS AND THE
78.8 PERCENT PROBLEM

Even the best-designed command relationships and the most elegant architectures will fail if human performance cannot keep pace. The fact that 78.8 percent of aviation mishaps have human factors components underscores the magnitude of the challenge. Digital tools, data feeds, and spectrum-dependent capabilities only add value if crews can integrate them into mission execution without becoming overwhelmed.

Major Moss's trajectory illustrates the kind of cognitive rewiring required for impact forces, treating information and the spectrum as co-equal to weapons rather than as background enablers.* This shift demands new patterns of training, different crew habits, and an instructional culture that values managing complexity as much as it values technical proficiency on individual systems. It also demands leaders who recognize that cognitive load is a design variable, not an afterthought, in how they build missions and configure networks.

Here, Hedelund's observation about the human element is decisive. In Steel Knight, he saw tents and buildings full of Marines who were engaged and hungry to learn, and he rightly judged that this enthusiasm mattered more than any single piece of kit. A force eager to learn will experiment, adapt, and find ways to exploit new tools faster than doctrine can be updated. That learning hunger is a core human factor advantage for impact forces operating under conditions of chaos and rapid change.

CULTURE AS THE DECISIVE ENABLER

At the end of the day, Hedelund's verdict on impact operations is blunt: you cannot do this with magic, and there is no voodoo solution. Success rests on the disciplined translation of concepts into capabilities, honest assessment of gaps, and sustained commitment to difficult organizational and cultural change. The Marine Corps already possesses platforms, digital interoperability, and the basic architecture of kill webs able to connect sensors and shooters across domains.

The unresolved question is whether the institution can build the command relationships, delegation authorities, and trust patterns needed to exploit those capabilities under the pressures of a contested spectrum and compressed decision timelines.

Steel Knight offered encouraging signs: Marines eager to learn, commanders willing to experiment, and tactical execution that met

---

* See the interview in chapter six in part two of this book.

very high standards. Those qualities suggest that the cultural foundation for impact forces is already in place, even if the institutional structures have not yet fully caught up.

Whether the Corps can evolve those structures fast enough to match its operational concepts will determine if it truly becomes the impact force that twenty-first-century strategic competition requires.

PART TWO
THE STEEL KNIGHT
25 INTERVIEWS

Part two shifts from analytical narrative to interview-based evidence, using conversations with Marine and joint practitioners to illuminate how the impact force is actually being built, tested, and contested in real organizations. It treats Steel Knight 2025 and associated initiatives as a laboratory observed from multiple vantage points, aviation, logistics, command and control, force design, and coalition engagement, allowing operators and commanders to articulate in their own terms what works, what breaks, and what still needs to change.

The opening interviews focus on I MEF and 3rd Marine Aircraft Wing leadership explaining why Steel Knight 2025 was structured as a campaign rather than a set of disconnected events. They describe the deliberate choice to abandon linear force closure, force units into distributed entry operations, and build a scenario that blended embassy reinforcement, noncombatant evacuation, distributed fires, contested logistics, and Marine Rotational Force–Darwin certification into one continuous problem set. Senior leaders stress that the exercise was designed to test whether Marines could function as an impact force from the first hours of a crisis, rather than merely validate existing procedures. Their commentary underscores a growing recognition that campaign design not just platform mix drives whether the impact force concept is real or rhetorical.

Subsequent interviews go down in echelon to regimental, battalion, and squadron-level commanders, who describe the practical challenges of executing distributed operations with timestamped nodes and constrained logistics. Ground commanders recount how the requirement to keep nodes moving, manage signatures, and maintain connectivity forced changes in task organization, load planning, and risk tolerance.

They highlight the tension between the need for redundancy. more equipment, more communications paths, and the need for smaller, lighter, less targetable footprints. These voices make clear that the physics of sustainment and signature management are reshaping what "forward presence" means on the ground.

A major strand in part two centers on digital interoperability and kill web integration from the perspective of communicators, air C2 specialists, and digitally enabled squadrons. Interviews with G-6 and MACG-38 personnel explain the "three ways to win" approach to communications design, the move from single exquisite links to an ecosystem of classified, commercial, and unclassified pathways, and the practical difficulty of balancing bandwidth, security, and survivability. They describe how tools like MAGTABs, gateway architectures, and new air operations center concepts are being used to push relevant data to the edge, while acknowledging that every added link complicates emission control and increases exposure to electronic attack.

Rotary-wing and unmanned aviation practitioners provide another layer, describing how platforms like the H-1 and MQ-9 are being transformed into essential nodes of the kill web. Squadron leaders outline how full digital interoperability allows helicopters once seen as vulnerable and legacy to act as low-altitude command-and-control extensions, pulling in F-35 and unmanned sensor feeds and pushing targeting data back into the joint network. VMU personnel discuss the MQ-9's evolving role as a strategic sensor and airborne relay, particularly in the Indo-Pacific, and the implications of new payloads and SATCOM options for persistent sensing and command support. These interviews collectively show aviation becoming less about discrete sorties and more about an integrated, distributed sensing and decision network.

Part two also includes extensive discussions with logistics commanders and staff who frame sustainment as the decisive constraint on impact force operations. They describe the difficulty of supporting constantly moving, widely dispersed nodes when every refueling point, ammunition cache, and maintenance site is targetable. Interviews emphasize dependence on joint and interagency partners, particularly Air Force airlift, and underscore the need for new concepts such as dynamic resupply, modular support packages, and closer integration of predictive maintenance tools that can reduce the logistics burden. These perspectives make clear that without a logistics system designed for chaos management, the impact force remains fragile.

A recurring theme in the interviews is command culture and authorities. Company and battalion commanders recount situations where junior leaders possessed detailed, high-quality situational awareness via modern C5ISR networks and data feeds but lacked the authority to act at the tempo the situation demanded. Senior leaders, in turn, describe the institutional pull toward centralization especially when connectivity is good and the difficulty of accepting that an impact force requires structural and cultural changes to give dispersed nodes more freedom of action. These conversations highlight that mission command in a kill web environment is still a work in progress, not a settled practice.

Part two also incorporates allied and partner perspectives, particularly from Australia and other Indo-Pacific observers who watched Steel Knight and related activities. Their commentary underscores how U.S. experimentation with impact force concepts is viewed abroad, where partners are grappling with similar challenges of contested logistics, persistent surveillance, and rapid adaptation. They point to the value of using major U.S. exercises as venues for professional military education, concept development, and alignment of operational expectations across coalitions.

Across these interviews, the war in Ukraine surfaces repeatedly as a reference point for adaptation speed, improvisation, and the integration of commercial technologies into high-intensity conflict. Practitioners caution against assuming that adaptation speed is uniquely or permanently a U.S. advantage and argue that exercises like Steel Knight must be treated as genuine opportunities for experimentation rather than scripted proofs of concept. They emphasize learning curves, unintended consequences, and the need to adjust doctrine, training, and acquisition priorities based on observed performance rather than preconceived narratives.

Part two, taken as a whole, functions as a multi-voiced operational audit of the impact force idea. It shows how senior leaders, staff officers, and tactical-level practitioners interpret the same problems through different lenses, and where their assessments converge: distributed operations can generate disproportionate effects through kill web integration, but only if logistics, digital architectures, and command culture are redesigned for chaos rather than crisis.

The interview material grounds the book's earlier arguments in concrete experiences, revealing both the progress already made and the unresolved tensions that will shape whether the emerging impact force can succeed in a future fight.

CHAPTER 1
OVERVIEW ON
STEEL KNIGHT 2025

Steel Knight 2025 is the Marine Corps' primary venue for testing what a stand-in, chaos-management force actually looks like at scale, rather than on PowerPoint. Built as a campaign of linked problems, embassy reinforcement, NEO, distributed fires, contested logistics, rather than a single set-piece battle, it forces I MEF to fight as a dispersed, networked MAGTF from the opening moments of a crisis.

The Marine Corps has long understood that the character of warfare is shaped not merely by weapons systems or technological innovation, but by how forces are trained, organized, and conceptually prepared for the operational environment they will face.

Steel Knight 2025 was conducted from December 1-14, 2025, across Southern California and the greater Southwest, represents a significant evolution in how I Marine Expeditionary Force (I MEF) prepares for the demands of Indo-Pacific contingencies.

This exercise transcends the traditional model of large-scale training events as discrete readiness checks, instead functioning as a "campaign laboratory" or a venue where operational concepts are tested, refined, and validated against the friction of realistic scenarios before they are deployed in actual theater operations.

THE SHIFTING PARADIGM: FROM FORCE CLOSURE TO DISTRIBUTED OPERATIONS

The design architecture of Steel Knight 2025 reflects a fundamental shift in Marine Corps readiness philosophy. Rather than organizing training around the assumption of linear force buildup and eventual force closure at secure bases, this iteration has been deliberately constructed to validate the ability to fight as a distributed, networked stand-in force from the opening moments of a crisis.[*]

This represents a doctrinal departure from the force-on-force engagements that characterized Cold War-era exercises and even many post-9/11 training events, where the implicit assumption was that U.S. forces would have time to aggregate combat power before engaging a near-peer adversary.[†]

The exercise brought together the core components of I MEF, 1st Marine Division, 3rd Marine Aircraft Wing, and 1st Marine Logistics Group, operating as an integrated Marine Air-Ground Task Force at division scale.

But rather than executing a single scripted battle, the exercise has been structured around a campaign-like sequence of problem sets: embassy reinforcement, noncombatant evacuation operations (NEO), dispersed fires and maneuver, and sustained logistics under threat.

The through-line connecting these vignettes is the requirement to synchronize sensors, decision-makers, and shooters across extended ranges and under compressed timelines, while sustaining dispersed ground and aviation elements from an expeditionary infrastructure rather than relying on the fixed-base sanctuary that characterized operations in Iraq and Afghanistan.

---

[*] Lorenzo Meigs, "Camp Pendleton Marines Commence Wargames for the Future Fight," Defense Visual Information Distribution Service, news release, December 3, 2025, https://www.dvidshub.net/news/552507/camp-pendleton-marines-commence-wargames-future-fight.

[†] United States Marine Corps. *A Concept for Stand-in Forces.* Headquarters, U.S. Marine Corps, Dec. 2021, https://www.hqmc.marines.mil/Portals/142/Users/183/35/4535/211201_A%20Concept%20for%20Stand-In%20Forces.pdf.

This approach mirrors what military historians have observed in other periods of doctrinal transition most notably the interwar period when the Marine Corps developed amphibious doctrine not through abstract theorizing alone, but through iterative experimentation at exercises like Fleet Landing Exercises (FLEXs).* Steel Knight 2025 serves a similar function for the stand-in force concept: translating strategic guidance into operational practice through repeated rehearsal under realistic constraints.

THE MRF-D CERTIFICATION IMPERATIVE: LINKING TRAINING TO OPERATIONAL DEPLOYMENT

Steel Knight 2025 is a dual function exercise as both a MEF-level training event and the formal certification venue for the regiment deploying as Marine Rotational Force – Darwin (MRF-D). Rather than treating certification as a discrete pre-deployment exercise separate from higher-level training, Steel Knight 2025 integrates the regimental certification process into a MEF-level campaign rehearsal, ensuring that the deploying unit validates its capabilities within the broader operational architecture it will support in theater.

The embassy reinforcement scenario at Camp Pendleton serves as the most visible manifestation of this integrated approach. The 5th Marines and associated aviation and logistics enablers rehearsed rapid reinforcement of a U.S. embassy under pressure, integrating MV-22 Ospreys and CH-53E Super Stallions to deliver a reinforced security element and the logistics backbone necessary to stabilize the situation and, if necessary, support noncombatant evacuation. The training environment incorporated urban terrain, role-players, and complex injects ranging from crowd control to casualty care and information operations, forcing commanders to manage simultaneous tactical, political, and humanitarian dimensions, precisely the kind of ambiguous crisis

---

* Trent Hone, "Amphibious Doctrine's Evolution in the Pacific," *Proceedings* 149, no. 6 (June 2023), https://www.usni.org/magazines/proceedings/2023/june/amphibious-doctrines-evolution-pacific.

scenarios that characterize actual embassy reinforcement and NEO missions in contested regions.[*]

This scenario is not a generic crisis-response drill but a tailored proof of concept: can this specific regiment, with this aviation and logistics package, function as a credible, rapidly deployable stand-in force capable of responding on short notice to crises affecting U.S. missions and citizens across the vast Indo-Pacific theater?

The answer to that question has direct operational consequences, as the certified regiment will deploy to Darwin as part of the forward-postured force architecture designed to provide regional combatant commanders with responsive options below the threshold of major combat operations.

DISTRIBUTED AVIATION ARCHITECTURE: SPOKES, FARPS, AND THE KILL WEB

Perhaps the most operationally significant innovation demonstrated during Steel Knight 2025 has been 3rd Marine Aircraft Wing's experimentation with a distributed aviation support architecture built around expeditionary "spokes" and forward arming and refueling points (FARPs).[†] During the exercise, 3rd MAW units established aviation spokes at Camp Pendleton and a FARP at Sacramento Mather Airport, approximately 400 nautical miles from Pendleton, demonstrating the wing's ability to project fuel, ordnance, maintenance, and command and control well beyond its main air stations.

These nodes allowed assault support and attack platforms to operate within striking distance of notional adversary targets while remaining supported from an agile, relocatable logistics network rather than a single, easily targeted base. This construct operationalizes the

---

[*] Robbin Laird, "The Embassy Reinforcement Exercise within Steel Knight 2025: The Osprey–Super Stallion Tandem and the Insertion Force," *Second Line of Defense*, December 2025, https://sldinfo.com/2025/12/the-embassy-reinforcement-exercise-within-steel-knight-2025-the-osprey-super-stallion-tandem-and-the-insertion-force/.

[†] Cpl. Nikolas Mascroft, *Steel Knight 25 B-Roll: 3rd MAW Establishes Spoke*, video, 1:51, December 2, 2025, Defense Visual Information Distribution Service, https://www.dvidshub.net/video/988461/steel-knight-25-b-roll-3rd-maw-establishes-spoke

Corps' kill-web logic: instead of viewing each airfield as a standalone hub, 3rd MAW is treating spokes and FARPs as interlocking nodes that can sustain aircraft, share data, and support fires across a widely dispersed battlespace.

In practical terms, this means tying together support squadrons, Marine Air Control Group 38 (MACG-38) air command and control detachments, and the ground combat element so that helicopters and tiltrotors supporting embassy reinforcement or other crisis-response tasks can be retasked, refueled, and rearmed through multiple routes. For a future Indo-Pacific campaign, this capability represents the difference between an aviation wing that must operate from a small number of vulnerable "big bases" and one that can rapidly establish and disestablish a web of survivable sustainment sites across a theater of operations.

During Steel Knight 25, Marines from Marine Wing Support Squadron 372, Marine Air Control Group 38, 3rd Marine Aircraft Wing, established and operated a Forward Arming and Refueling Point at Sacramento Mather Airport, California, on 8 December 2025."[*]

This distributed architecture also addresses a fundamental vulnerability in modern military operations: the concentration of critical capabilities at fixed, targetable locations.[†] By dispersing aviation support across multiple expeditionary nodes, 3rd MAW complicates adversary targeting and increases the resilience of the overall force. If one spoke or FARP is neutralized, the network can adapt and reconstitute support through alternate nodes, a form of operational redundancy that is essential when facing adversaries with sophisticated intelligence, surveillance, and reconnaissance capabilities and long-range precision strike weapons.

---

[*] Sgt. Brian A. Stippey, *Steel Knight 25: Diamondbacks Establish Mather FARP* [photograph], December 8, 2025, Defense Visual Information Distribution Service, Photo ID 9430867, https://www.dvidshub.net/image/9430867/steel-knight-25-diamondbacks-establish-mather-farp.

[†] Sgt. Brian A. Stippey, "Steel Knight 25: 'Support Within Striking Distance' | Diamondbacks Deliver Long-Range Aviation Support," *Defense Visual Information Distribution Service (DVIDS)*, December 10, 2025, Story ID 553791, https://www.dvidshub.net/news/553791/steel-knight-25-support-within-striking-distance-diamondbacks-deliver-long-range-aviation-support.

THE COALITION DIMENSION: STEEL KNIGHT AS THEATER ENGAGEMENT

Steel Knight 2025 has also retained and expanded its multinational dimension, with International Observation Programs and partner-nation delegations present at Camp Pendleton and other venues. These observers use the exercise as a window into how I MEF trains for high-end conflict and crisis response in a contested maritime region. While embassy reinforcement and NEO scenarios are familiar tasks to many regional partners, the way I MEF integrates these missions within a larger campaign construct, complete with distributed aviation support, integrated fires, and sustained medical and logistics operations, offers a concrete model of how a stand-in MAGTF might operate alongside allies.

This coalition-building aspect should not be underestimated. In an era where strategic competition is as much about building and maintaining alliances as it is about combat capabilities, exercises like Steel Knight serve a theater-shaping function by demonstrating U.S. operational concepts to partners and by creating opportunities for professional military education and relationship-building. The presence of international observers transforms Steel Knight from a purely national training event into a form of security cooperation, where the exercise itself becomes a medium for communicating operational approaches and building shared understanding of how U.S. and partner forces might operate together in a crisis.

CONCLUSION: CAMPAIGN READINESS IN THE INDO-PACIFIC CONTEXT

The public narrative surrounding Steel Knight 2025 positions it as more than an annual readiness check; it is portrayed as an iterative step in building a Marine Expeditionary Force that can function as a theater-shaping stand-in force, one that can reinforce an embassy today, support an ally's coastal defense tomorrow, and feed targeting and sustainment into a joint and combined kill web over time.

The embassy event at Camp Pendleton's training village and the spoke/FARP construct at Pendleton and Mather provide tangible, observable anchors for this broader strategic concept, bridging the abstract language of kill webs and stand-in forces with the concrete realities of operational execution.

What emerges from Steel Knight 2025 is a vision of Marine Corps readiness that is fundamentally different from the garrison-to-deployment model that dominated the service for decades. Instead of viewing training as preparation for eventual combat operations, this exercise treats training as continuous campaign rehearsal, a way of maintaining a force that is perpetually ready to operate as a distributed network rather than a concentrated formation.

This approach acknowledges the reality that in the Indo-Pacific theater, crisis response and warfighting exist on a continuum rather than as discrete phases, and that the force must be equally prepared for ambiguous below-threshold operations and high-intensity combat.

For strategic analysts and defense planners, Steel Knight 2025 offers important insights into how the Marine Corps is operationalizing the stand-in force concept and preparing for the operational demands of great power competition.

The exercise demonstrated that the service is not simply proclaiming new concepts but systematically validating them through realistic, stressful training that integrates all elements of the Marine Air-Ground Task Force.

As the certified regiment deploys to Darwin and as subsequent rotations prepare for similar forward deployment, the lessons and validated procedures from Steel Knight 2025 will shape how the Marine Corps postured across the Indo-Pacific executes its mission as a crisis-response and theater-denial force.

The success or failure of this operational model will ultimately be judged not in training areas but in actual contingencies, but Steel Knight 2025 represents an essential step in building the institutional competence and operational confidence necessary for that future mission.

CHAPTER 2
HOW 3RD MAW IS TRANSLATING STRATEGIC CONCEPTS INTO COMBAT-READY CAPABILITIES

3rd Marine Aircraft Wing sits at the center of the Corps' aviation-driven transformation, turning ideas like stand-in forces, DAO, and kill webs into concrete sortie generation, basing schemes, and C2 architectures. This chapter shows how 3rd MAW leadership interprets Force Design 2030 and chaos management not as a theory exercise, but as a demand to rewire training, maintenance, and deployment patterns to support a truly distributed fight. It sets up the subsequent chapters on specific groups and squadrons by explaining the Wing-level logic that ties their experiments together.

In a wide-ranging discussion at Marine Corps Air Station Miramar, Colonel Michael Perrottet, the G-3 (Operations) officer for 3rd Marine Aircraft Wing, offered a perspective on force transformation that moves beyond theoretical debates about Force Design 2030. His view emphasizes practical adaptation over doctrinal purity, revealing how the wing is actively reshaping Marine Corps aviation capabilities for an uncertain future.

Colonel Perrottet frames Force Design 2030 not as a fixed blueprint but as an evolving experiment. The key insight is that Force Design's true value lies not in its specific prescriptions but in the framework it provides for continuous innovation. "We are excellent at being the guardians of the perennial task of creative destruction," Perrottet observed. "We constantly build something up, and then we are the ones to tear that down and progress it forward."

Perrottet articulates 3rd MAW's operational approach through three distinct levers that commanders can manipulate:

- Tasking: The mission sets and their prioritization across different echelons, adjusted to meet evolving strategic requirements.
- Capability: The range of aviation assets from fifth-generation fighters to assault support, each with unique attributes that enable different operational approaches.
- Capacity: The depth, breadth, training, and readiness of forces, essentially the quantitative dimension of what the wing can sustain over time.

"Those are the three things that we're constantly trying to meet," Perrottet explained. This framework provides a practical way to think about force employment that transcends abstract debates about force structure.

When pressed about what enables the Marine Corps to adapt so effectively, Perrottet returned to fundamentals. "The core element is the people."

This emphasis on human capital reflects a deeper truth about military innovation: technology alone changes nothing. "You can mix and match our people in any way, shape or form, we're going to be quick to adapt," Perrottet said. The organization must co-evolve with its technology, not simply bolt new systems onto static structures.

This becomes particularly important as 3rd MAW integrates autonomous systems, both maritime and airborne. The wing's location on the West Coast positions it uniquely among combat units, surrounded by companies developing cutting-edge autonomous capabilities from San Francisco to San Diego. "

The evolution of Exercise Steel Knight illustrates how 3rd MAW is expanding its operational aperture. What began in 1991 as a small tank exercise has grown into a massive joint operation spanning from Washington State to Yuma, Arizona, involving Marine Corps, Navy, and Air Force elements.

"In 2023, we welcomed ESG-3, actually Third Fleet writ large, to participate," Perrottet explained. "Now in 2025 we've got the Marine Corps, the Navy and the Air Force." This expansion reflects what he terms "N3I" or naval integration, interdependence, and interoperability extending across service boundaries.

The exercise serves three essential functions: warfighting readiness, force generation, and modernization through experimentation. "I don't think it's either modernization experimentation or the development of that warfighting readiness," Perrottet said. "I think it's all three of those."

Critically, Steel Knight provides the venue for controlled failure. "We have to be willing to fail in a training and rehearsal environment," Perrottet emphasized. "If we're not willing to fail, then we will avoid those opportunities to fail. And ultimately, all you're going to do is lie, hide and fake the funk."

Perrottet sees the Marine Corps' expeditionary character as its defining strength in an era of distributed operations. "One thing we are able to do well are expeditionary operations. We get forward, we persist, we survive. We're lethal," he said. "I think we can do that in a sustained way."

This expeditionary ethos extends to how the wing thinks about autonomous systems and emerging technologies. Rather than waiting for perfect solutions, 3rd MAW incorporates what works now. The ability to rapidly integrate contractor systems during exercises, evaluate their performance, and make fielding recommendations represents a combat development cycle measured in months rather than years.

Perrottet articulated five critical drivers of effective change management: vision, skills, incentives, resources, and implementation. The absence of any one element creates specific organizational pathologies:

- Without vision: confusion.
- Without skills: anxiety.
- Without incentives: resistance.

- Without resources: frustration.
- Without implementation: false starts.

"The piece about implementation is not enough," he emphasized. "I look at it as a continuous process." This perspective explains why Force Design remains contentious, implementation is ongoing, not complete, and the Marine Corps continues making course corrections based on operational reality.

Regarding the broader strategic picture, Perrottet acknowledged the uncertainty inherent in planning for conflict with a peer competitor. The challenge isn't just technical or operational for its epistemological. How well do we actually understand how a potential adversary would fight?

This uncertainty reinforces the importance of adaptability over rigid planning. In Perrottet's view, the Marine Corps serves as a probing force, able to operate in contested environments and generate the kind of operational knowledge that can only come from contact with the enemy. But this role demands the right mix of survivability, lethality, and mobility to ensure Marines can accomplish their mission and get out alive.

Colonel Perrottet's perspective on 3rd Marine Aircraft Wing's transformation offers a compelling alternative to the binary debates that often characterize discussions of Force Design 2030. Rather than defending or attacking a particular vision, he describes an organization engaged in continuous experimentation, learning, and adaptation.

The wing's approach emphasizes several key principles:

- Use what you have effectively rather than waiting for perfect solutions.
- Test concepts through realistic exercises that permit failure.
- Maintain the core competencies that make Marine aviation unique.
- Integrate new technologies as complements to existing capabilities.
- Keep people, their character and competence, at the center of transformation.

This pragmatic approach may lack the dramatic appeal of revolutionary manifestos, but it reflects how military organizations actually evolve in practice. By focusing on operational realities rather than abstract concepts, 3rd MAW is building the force that can fight tonight and tomorrow.

As Perrottet put it simply: "We want to ensure we win the wars and the battles. And so to do that, we're going to constantly reiterate and review and have our own internal OODA loops to self-correct." In an uncertain strategic environment, that capacity for continuous adaptation may matter more than any specific force structure or operational concept.

And if there is any lesson to be taken from the war in Ukraine, is that the Ukrainian ability to adapt rapidly has been at the core of their unanticipated success. Among the U.S. services the Marines, in my view, are the most likely element along with the SOF forces of being able to drive a similar innovation path.

TURNING DISTRIBUTED OPERATIONS INTO WARFIGHTING PRACTICE

Distributed operations only matter if they change how Marines close with the enemy, protect sustainment, and generate effects under pressure. Here I follow how I MEF and 3rd MAW used Steel Knight 2025 to move DAO from slideware to executable TTPs: dispersing forces, synchronizing sensors and shooters across distance, and sustaining combat power without the comfort of large, secure hubs. The focus is on where distribution created genuine warfighting advantage—and where it risked becoming dispersion for its own sake.

At the recent Steel Knight 2025 exercise conducted by 3rd Marine Aircraft Wing at MCAS Miramar, Lieutenant General (Retired) Hedelund served as a senior mentor, bringing decades of operational experience to observe and guide one of the Marine Corps' most significant training evolutions. His observations reveal both the promise and challenges facing the Corps as it transforms from Force Design 2030 concepts into practical operational capability.

THE MAGTF STAFF TRAINING PROGRAM (MSTP): PRESERVING INSTITUTIONAL KNOWLEDGE

Hedelund's role at Steel Knight exemplifies the Marine Corps' commitment to leveraging retired general officer expertise through the MAGTF Staff Training Program (MSTP) (MSTP). This program, which has existed for decades, provides a stable of experienced flag officers who assist Marine Expeditionary Force (MEF) commanders and staffs with training and certification exercises.

"It's not about making money," Hedelund explained. "For those of us that stick with it, it's more about what we can contribute. It's about staying connected to the service that we love, and more importantly, the people that we serve with inside the Marine Corps." This emphasis on genuine institutional commitment ensures that the mentor corps consists of officers who "care about what the future of the Marine Corps holds."

The program operates under strict guidelines following reforms that addressed earlier concerns about senior mentor activities. Today's program maintains a focused cadre of highly qualified experts who work within clearly defined parameters, ensuring both accountability and mission effectiveness.

CONTINUITY AMID TRANSFORMATION

One of Hedelund's most striking observations challenges the narrative that Force Design 2030 represents a complete break from Marine Corps history. While acknowledging the impact of new technology and capabilities, he emphasized fundamental continuity in Marine Corps operations.

"If you peel back the words and the semantics, there's a lot that I did in 1985 on my first deployment that I saw in Steel Knight," Hedelund noted. "Different tools, different capabilities, eye-watering technology that enables quicker decision making and more rapid deployment over longer distances. But the Combined Arms Force is still the secret sauce."

Whether organized as a Marine Air-Ground Task Force (MAGTF), Marine Littoral Unit (MLU), or another formation, the essential components remain: a command element, capable aviation, ground combat forces, and critically, logistics. "It's still essentially an air-ground team," Hedelund observed, though he expressed concern about potentially inadequate logistics capacity to support distributed operations.

This continuity matters because it provides templates for commanders grappling with new operational concepts. Marines aren't inventing distributed operations from scratch. They're adapting proven combined arms approaches with dramatically enhanced technological capabilities including the F-35, MV-22 Osprey, and the incoming CH-53K King Stallion.

THE COMMAND-AND-CONTROL CHALLENGE

Steel Knight 2025 revealed both impressive progress and persistent challenges in command and control for distributed operations. Hedelund observed 3rd MAW attempting to function as both force provider and operational headquarters from Miramar is a demanding dual role that exposed important questions about command relationships.

"The acknowledgement that nodes, communication nodes, command and control nodes, logistics nodes, all have some vulnerability no matter where they are in this world today," Hedelund explained. "How do we fight with the nodes that we have at the moment that we have them?"

The exercise highlighted complex questions about who actually commands air operations in a distributed Pacific scenario. When a Marine division becomes the lead forward element and eventually transitions to Joint Task Force status, who provides air component expertise and resources?

"What's the relationship between the JFACC and the 3rd MAW commander, or the third MAW commander's representative in theater if you put, say, a MAW forward out into the Philippines?" Hedelund asked. This question gets at the heart of joint operations in distributed environments, ensuring clear command relationships while enabling the rapid decision-making that modern combat demands.

TACTICAL EXCELLENCE, STRATEGIC QUESTIONS

While command and control challenges persist, Hedelund expressed strong confidence in tactical execution. "The mission execution was really well done," he observed. "I was really impressed with the packages put together to address some of the threats. They did impressive work against a realistic threat. Nobody is worried about the actual tactical execution. Marines will be Marines."

The exercise demonstrated effective use of assault support packages and combined arms operations against realistic threats. Marines at the tactical level showed they could employ distributed forces effectively, integrating aviation and ground elements to accomplish specific missions.

However, Hedelund's concern centered on ensuring these tactical capabilities serve coherent strategic purposes. The exercise tested various force projection scenarios and different combinations of capabilities within a kill web framework, but questions remained about how these tactical operations would integrate into broader campaign plans.

HUB, SPOKE, AND NODE OPERATIONS

3rd MAW worked hard on developing hub, spoke, and node concepts for distributed operations, but Hedelund identified areas requiring further maturation. "You have to make sure that everybody out there understands what a hub is and what it can provide, and likely what it cannot provide, and how static it is, or how permanent or semi-permanent it is. Same thing with the spoke. And then down to the node."

Current nodes remain "a little bit too heavy" in Hedelund's assessment, partially because the Corps hasn't fully exploited F-35 capabilities. The challenge involves determining precisely what each node should accomplish and for how long, then designing appropriate support structures.

Hedelund proposed an important operational concept: putting timestamps on nodes. "You almost have to tell whoever's in charge of that node you need to assume that you're coming out at time X on date X, unless you're told otherwise and not the other way around where you're asking 'We're about to get pummeled. Can I come home?'"

This approach treats nodes as having inherent shelf lives which are measured in hours or days depending on their assigned tasks. "They've got a shelf life, and that shelf life may be hours or maybe a couple of days, depending on what task it's been told to accomplish," Hedelund explained. Elements should constantly move, with forces that inserted a node already en route to establish the next one while the current node completes its mission.

This concept addresses signature management or the reality that forces firing missiles from a location for extended periods will be detected and targeted. Constant movement becomes essential to survival and effectiveness in contested environments.

THE LOGISTICS IMPERATIVE

Throughout our conversation, both Hedelund and I returned repeatedly to logistics as the critical enabler and limiting factor of distributed operations. "I'm keeping my eye on the logistics piece of this, because that is the real key or cornerstone of any success," Hedelund emphasized. "The distances we are talking about require a capable logistics system and organization."

Contested logistics presents challenges without perfect solutions, rather only paths to attenuating problems. The CH-53K King Stallion "couldn't come fast enough" in this context, and maritime autonomous systems offer promise for sustaining distributed nodes. Yet fundamental questions remain about what shipping the Navy will provide for Marine Corps operations.

The Air Force's 130 aircraft proved crucial to Steel Knight's success. "If we didn't have Air Force 130s in this exercise, things would have looked a lot different," Hedelund observed. This dependency highlights capability gaps facing the Marine Corps. In my view, the USMC doesn't have enough KC-130Js, MV-22 Ospreys, CH-53Ks, or F-35s to fully execute its distributed operations concept without joint support.

The CH-53K could potentially assume many KC-130J missions, but only if acquired in sufficient numbers. "You cannot do this operation with magic," Hedelund concluded. "You might want to, but it's not going to happen. No voodoo here."

DELEGATION AND DECISION AUTHORITY

Both Hedelund and I agreed that distributed operations demand unprecedented delegation of decision authority. The technology and lethality available to smaller units has increased exponentially, but organizational structures haven't fully adapted to enable these capabilities.

In my view, if you're projecting force and distributing that force, you're going to have to delegate decision making in ways we've never done historically. The challenge facing INDOPACOM involves more than Marine Corps force structure. It requires rethinking how area commanders task subordinate units and what authorities those units need to exploit fleeting opportunities.

Modern C5ISR and networking capabilities mean that relatively small Marine units possess situational awareness and strike capabilities that once required much larger formations. An F-35, an Osprey, and a CH-53K fundamentally change what a Marine unit can accomplish independently. Add maritime autonomous systems and eventually Collaborative Combat Aircraft, and small distributed nodes possess firepower equivalent to much larger forces a decade ago.

But capability without clear tasking and appropriate authority becomes wasted potential. As Hedelund noted, "You're going to be hoping that your commander is giving you clear guidance, and you're going to be executing that guidance, yet may not be able to get feedback right away on whether or not what you did was a good thing or a bad thing."

MISSION COMMAND IN PRACTICE

The Air Force struggles with similar challenges, Hedelund observed from his work at the LeMay Center. "They have embraced mission command. They understand the 'what' but are still figuring out the 'how.' It's like the dog that caught the tailpipe of the car. Now what?" They understand distributed operations will be necessary in the Pacific, "but yet they struggle to let go of the centralized control piece of it."

This tension between centralized control and distributed execution isn't unique to any service. Operating in contested electromagnetic environments where a 32-second radio transmission could bring retribution within two minutes requires fundamentally different approaches to command and control.

Steel Knight 25 experimented with low-signature communications and rapid decision cycles, but questions remain about how much authority to delegate and how to maintain unity of effort across distributed nodes operating with limited communications.

THE PATH FORWARD

Despite challenges, Hedelund came away impressed with the exercise's professionalism and the participants' commitment to learning. "In every tent or building that we ended up in during the exercise, everybody was engaged. They were eager and hungry to learn more and to practice these things that people have been talking about. There's nothing more exciting than people who are hungry to learn."

Major General Wellons willingness to embrace outside perspectives exemplified the mature command climate Hedelund praised. "There are some commanders out there that aren't comfortable with people from outside coming in and trying to assist," Hedelund noted. "That willingness shows a real maturity level in our commanders."

Steel Knight 2025 demonstrated that the Marine Corps is translating distributed operations concepts into practical capability. Tactical execution remains strong, and Marines at all levels show commitment to mastering new operational approaches. The exercise tested multiple force distribution scenarios and integrated kill web operations, providing valuable insights for continued development.

However, significant work remains in three critical areas.

- First, command and control structures must evolve to match distributed operations' tempo and decision-making requirements.
- Second, logistics capabilities need substantial enhancement through additional platforms like the CH-53K and integration of maritime autonomous systems.
- Third, and perhaps most importantly, the Corps needs clearer strategic guidance on how distributed forces will be employed: what specific tasks nodes will accomplish, for how long, and in support of what larger campaign objectives.

As Hedelund concluded, "You cannot do this operation with magic." Success requires continued hard work translating concepts into capabilities, honest assessment of gaps and shortfalls, and sustained commitment to the difficult organizational and cultural changes that distributed operations demand.

The Marine Corps that deployed Hedelund in 1985 and the Marine Corps conducting Steel Knight 2025 share fundamental DNA: both are combined arms forces built around aviation-ground integration. But the tools available today enable operations at scales and speeds previously impossible. The challenge now is ensuring institutional structures, command relationships, and logistical capabilities match the revolutionary potential of these new operational concepts.

CHAPTER 4

ASSAULT SUPPORT TRANSFORMATION: A DISCUSSION WITH MAG-16 COMMANDERS

Squadron (HMH) 465, Marine Aircraft Group 16, 3rd Marine Aircraft Wing, transports Marines with 1st Battalion, 5th Marine Regiment, 1st Marine Division, during a simulated embassy reinforcement as part of Exercise Steel Knight 25 at Marine Corps Base Camp Pendleton, California, Dec. 5, 2025. Marine Corps photo by Sgt. Kyle Chan.

Assault support is the mobility backbone of the distributed force, and MAG-16's commanders describe how they are reinventing that role for a world of long-range fires and contested logistics. In Steel Knight 2025, their squadrons used spokes, FARPs, and expeditionary nodes to keep Marines moving, supplied, and connected across a wide battlespace rather than shuttling between a few big bases. This chapter captures how assault support leaders think about survivability, sustainment, and digital integration as coequal design drivers, not afterthoughts

Marine Aircraft Group 16 commanders, gathered during exercise Steel Knight at Marine Corps Air Station Miramar, offered insights into how assault support aviation is adapting to contested environments and distributed operations. Their collective observations reveal both the remarkable evolution of Marine Corps aviation capabilities and the substantial challenges that lie ahead.

Marine Aircraft Group 16 serves as the Marine Corps' premier assault support organization, providing the MAGTF commander with the ability to move combat troops and supplies day or night, in all weather conditions, across the full range of military operations. This mission is enabled by four MV-22B squadrons and four CH-53E squadrons, supported by an aviation logistics squadron that maintains and supplies these critical platforms.

Col Nate Storm, the group commander and a veteran CH-53E pilot who previously commanded HMH-462, explained that while MAG-16 headquarters was not directly involved in Steel Knight as a headquarters element, several squadrons were actively providing assault support for the exercise. This participation reflected the ongoing refinement of concepts that have been developing since the Commandant's planning guidance redirected the Marine Corps toward naval integration and distributed operations.

EVOLUTION SINCE THE
FORCE DESIGN SHIFT

The transformation of assault support aviation has accelerated dramatically since 2018, when the focus shifted toward great power competition. One commander traced this evolution back to the Commandant's guidance emphasizing a return to "soldiers from the sea" and renewed naval integration. After twenty years operating in the deserts of Iraq and Afghanistan, Marine aviation is now refocusing on littoral operations.

The biggest change that we're seeing right now is what we're trying to execute during Steel Knight — distributed air operations," one senior officer explained. "The focus of individual squadrons and bigger exercises like this are specifically focused on that ability to provide the ground combat element its ability to move around the battle space in a distributed function."

This shift represents more than tactical adjustment. It reflects fundamental rethinking of how assault support platforms operate across extended ranges in contested environments where traditional support structures may not be available or may become targets themselves.

An MV-22B Osprey tiltrotor aircraft prepares to land as U.S. Marines with 1st Battalion, 5th Marine Regiment, 1st Marine Division, wait to board during Exercise Steel Knight 25 at Marine Corps Base Camp Pendleton, California, Dec. 7, 2025. This exercise will certify 5th Marines for Marine Rotational Force–Darwin, a six-month forward deployment in Australia that strengthens combined interoperability with the Australian Defence Force and provides rapid crisis-response options for the joint force across the Indo-Pacific. U.S. Marine Corps photo by Lance Cpl. Caleb Goodwin.

THE V-22 CONTRIBUTION

The Osprey's integration into Marine aviation provides perhaps the clearest example of how platforms have outpaced conceptual thinking. One commander who attended Weapons and Tactics Instructor course in 2008 — part of only the second WTI class where V-22 students participated as actual class members rather than just support — recalled the intense learning process.

"Every time we would get done with an event, the instructors would be back writing notes, because they were still trying to figure out how we were going to employ that aircraft," he remembered. "It was already pretty apparent that it wasn't a replacement for the CH-46, like it had been bandied about. It's not. It has so much different capabilities. If you're just going to say that it's going to be a replacement for a CH-46, you're missing the point."

Nearly twenty years later, that assessment has proven prophetic. The V-22 performs a vastly expanded mission set, traveling significantly farther and faster than its predecessor. As one commander noted, the aircraft's capabilities enable operations that simply would not have been possible with the CH-46.

However, this capability expansion brings its own challenges. Demand for V-22 capabilities continues growing across the fleet, with new missions emerging regularly. The recent addition of sonobuoy deployment capability and the potential for deploying torpedoes from V-22s demonstrates how the platform continues finding new roles.

One significant development which I introduced into the conversation was the nacelle improvement program, which has reportedly exceeded expectations in terms of readiness improvements and maintenance reduction. Reportedly, the Air Force has seen a dramatic reduction in maintenance burden translates directly to increased readiness and the ability to redirect maintenance hours toward improving other aspects of aircraft capability. Applied to the Marines would mean more birds available to the operational commanders.

RANGES, REFUELING AND THE REALITY OF DISTRIBUTED OPERATIONS

Multiple commanders emphasized that modern assault support operations increasingly occur at ranges that exceed traditional close air support capabilities. V-22s and CH-53s routinely operate beyond where rotary-wing escorts can follow, pushing into areas where establishing forward arming and refueling points for escort aircraft would compromise operational security.

"We're continually working at longer ranges than we have in the past," one officer explained. "Part of that's driven by distributed aviation operations, part of it's just the nature of the threat. What we're seeing is a trend more towards disaggregated units, where the ground force might be part of the stand-in force, but the aviation is going to be separated."

This separation creates new integration requirements between strike and assault support communities. Rather than sequential operations — strike one day, assault two days later — current planning emphasizes synchronized operations where strike packages set conditions for assault packages in rapid succession, followed by planned retrograde operations.

The tyranny of distance creates another persistent challenge: aerial refueling. One relatively new squadron commander described a mission flying to the Philippines where his aircraft came within 500 pounds of a mandatory plug-in with the tanker.

"It's a little unnerving launching, knowing that your tanker may or may not show up," he admitted. "I don't know if we train to that as much as we would like to."

The concern extends beyond simply whether tankers will be available. In contested environments, tanker support becomes even more problematic. KC-130s may be tasked with other missions, and Air Force tankers operating from second island chain locations may have their own priorities. For assault support aircraft operating at extreme range, failed tanker rendezvous could prove problematic.

I introduced the notion that one interesting technological development being worked by industry is autonomous air systems to provide for tanking support.

THE DISTRIBUTED OPERATIONS REALITY CHECK

The electromagnetic signature problem presents a significant challenge. Current systems used for communication and coordination emit powerful electromagnetic signatures that work well in exercises and permissive environments but would prove disastrous in contested operations.

The challenge of being able to operate in electromagnetic silence (EMCON) is a significant one and the Marines and the joint force are working this challenge very hard.

THE MAINTENANCE FOOTPRINT CHALLENGE

Every commander emphasized that distributed operations require accompanying maintenance packages. The question isn't whether to send maintainers forward — that's essential — but rather determining the right size and structure of maintenance detachments and, critically, which spare parts to pack.

"We've cracked the nut on the people, the size and structure of the types of people that go with it," one officer assessed. "I don't think we've totally cracked the nut yet on the parts pack-up."

The complexity of modern aircraft makes predicting required spare parts extremely difficult. Even sophisticated platforms like the F-35, which feature advanced diagnostics, still require manual verification of fault indications. The tolerance thresholds in current systems aren't refined enough to provide absolute confidence that reported faults are genuine.

The challenge becomes more acute when forward deployed. Units essentially gamble on having the right components. If they guessed wrong, aircraft remain grounded until parts arrive - if they can arrive at all in contested environments.

PREDICTIVE MAINTENANCE: PROMISE AND PROBLEMS

The CH-53K represents a potential solution with its advanced predictive maintenance capabilities. As one commander explained, the ability to know when aircraft will require specific maintenance and plan accordingly could transform distributed operations.

"If you're in distributed aviation operations with targeting cycles, if you can say 'I know when this aircraft is going to need to have a certain thing, or how much life I've got on a particular component,' and I can actually schedule when that aircraft can be down, have the parts and the maintainers routed to that location, affect the maintenance, and move on — that's where the money is going to be made," he said.

However, even with predictive maintenance, challenges remain. The Navy's supply system hasn't matured sufficiently to exploit predictive maintenance capabilities fully. Parts availability remains a persistent problem across the service.

ADDITIVE MANUFACTURING: THE LONG ROAD AHEAD

One officer with extensive experience in advanced manufacturing offered sobering assessment of 3D printing's current limitations for aviation maintenance. While additive manufacturing shows promise, significant hurdles remain.

Parts produced through additive manufacturing must meet or exceed original equipment manufacturer specifications, but obtaining the proprietary technical data necessary to produce them remains difficult. The timeline from requesting approval to produce a part until it's ready ranges from eight to eighteen months.

Currently, additive manufacturing works better for ground combat elements with less stringent quality requirements. For aviation, until program offices mandate additive manufacturing for specific parts from the outset, adoption will proceed slowly.

FORWARD POSITIONING AND THE COMPETITIVE OPERATIONAL SPACE

One operational commander recently returned from a unit deployment program to the First Marine Aircraft Wing described successful long-range operations conducted entirely in EMCON. "I think the blocking and tackling of assault support largely stays the same," he reflected. "But we're looking into ways to extend the legs, because back when I was growing up as a pilot, we were not prioritizing aerial refueling or the long-range capabilities of the platform. We were very OEF/OIF-centric."

The experience highlighted both possibilities and limitations. Forward-leaning operations during competition phase before actual conflict allow positioning of logistics and support. The hub-spoke-node concept embraced by most Marine expeditionary units aims to create common operational pictures for logistics while pre-positioning supplies at multiple Global Prepositioning Network sites.

THE INTEGRATION IMPERATIVE

Multiple commanders stressed the growing importance of integration with strike aviation. Traditional assault support often relied on rotary-wing escorts, but future operations will increasingly require fixed-wing support.

This integration extends beyond simple escort to comprehensive packages where strike and assault support elements work in carefully coordinated sequences. Weapons and Tactics Instructor courses now emphasize these integrated evolutions, with strike packages setting conditions immediately before assault packages execute their missions.

A recurring theme throughout the discussion was the requirement to remain below enemy targeting thresholds while maintaining lethality. This proves particularly critical in the Indo-Pacific theater where peer adversaries possess sophisticated sensor networks and long-range strike capabilities.

THE LOGISTICS PACING FUNCTION

While concepts drive headlines, logistics ultimately determines what's operationally feasible. Multiple commanders emphasized this point with varying degrees of bluntness.

The group's focus on hub-spoke-node operations reflects recognition that pure disaggregation becomes unsustainable. Some consolidation of support proves necessary, but determining the right balance between disaggregation for survivability and consolidation for efficiency remains an ongoing challenge.

The emerging operational concept emphasizes short-duration forward deployments rather than sustained presence. This approach minimizes exposure while maximizing impact.

"I think it's more lucrative for us to be very short term, short duration," one squadron commander explained. "Practicing like that - you're going to get there with whatever flight of four or five or six V-22s on day one and maybe day two. Day three is the CH-53 logistical support. What do you do if it doesn't get there? I think we need to be doing more of that type of training operation."

THE PATH FORWARD

Throughout the discussion, several themes emerged consistently.

- First, platforms — particularly the V-22 — have capabilities that exceed current employment concepts.
- Second, distributed operations remain immature conceptually and especially in execution.
- Third, logistics and maintenance for distributed operations require fundamental rethinking.
- Fourth, integration between strike and assault support must continue deepening.

Yet capability without sustainable support structures and realistic command and control proves of limited utility. As these commanders make clear, assault support transformation requires more than acquiring better platforms. It demands new thinking about how to sustain, command, and employ those platforms in environments fundamentally different from those of the past two decades.

The conversations at MAG-16 during Steel Knight revealed professionals grappling honestly with these challenges. Their insights, grounded in operational experience and focused on practical execution rather than conceptual purity, provide essential perspective on where Marine assault support aviation stands and where it must go to remain relevant in an increasingly contested future.

Participants in the Discussion:

- Col Nate Storm, MAG-16 Commander
- LtCol Brian Spillane, HMH-462
- LtCol Michael Drozd, HMH-465
- LtCol Allen Whitlow, VMM-161
- LtCol Ryan O. Martin MALS-16
- Lt. Col. Justin K. Sing, MAG-16 Executive Officer

Ed Timperlake and I interviewed Lt Col Sing earlier when he was at MAWTS-1 and that interview can be found in our coauthored book on MAWTS-1.

CHAPTER 5

MAXIMIZING FLEXIBILITY: MAG-11'S APPROACH TO MULTI-DOMAIN AVIATION OPERATIONS

U.S. Marines with Marine Wing Support Squadron 372, Marine Aircraft Group 39, 3rd Marine Aircraft Wing, expeditiously load an AIM-120C Advanced Medium-Range Air-to-Air missile into a F-35C Lightning II assigned to Marine Fighter Attack Squadron (VMFA) 311, MAG-11, 3rd MAW, during exercise Steel Knight 25 at Southern California Logistics Airport California, Dec. 10, 2025. The Victorville node provided the infrastructure needed for rapid re-arming and refueling, allowing 3rd MAW aircraft to remain lethal and responsive in a simulated modern contested environment.(U.S. Marine Corps photo by Cpl. Renee Gray).

If MAG-16 provides the muscle of movement, MAG-11 brings a multi-domain strike and sensing brain that must flex across missions and theaters. During Steel Knight 2025, MAG-11 treated its aircraft, sensors, and data links as a configurable toolkit for supporting both crisis-response vignettes and higher-end combat problems. This chapter examines how the group thinks about flexibility in practice—re-tasking platforms across roles, managing signatures, and plugging into joint and coalition kill webs.

In a conversation conducted during Steel Knight 2025 at a simulated forward-deployed location, Colonel Jarrod DeVore, Commanding Officer of Marine Aircraft Group 11 (MAG-11) within 3rd Marine Aircraft Wing, provided an overview of how the Marine Corps is evolving its aviation capabilities to meet the challenges of distributed operations in contested environments.

MAG-11 is 3rd Marine Aircraft Wing's primary fixed-wing combat aircraft group, centered at MCAS Miramar. MAG-11's primary mission is to provide air support to Marine Air-Ground Task Force (MAGTF) commanders, including offensive air support, anti-air warfare, assault support, aerial reconnaissance, and terminal area control of aircraft. It also generates critical aviation logistics and significant aviation ground support functions to keep those fixed-wing forces sustained in combat.

Within 3rd MAW, MAG-11 is the main fixed-wing group responsible for training, deploying, and integrating F/A-18, F-35C, and KC-130 units, and for providing squadrons to Navy carrier air wings when required. The group can deploy its headquarters as a site command to run high-tempo sortie generation for the Marine Tactical Air Commander in support of I MEF and other combatant commanders.

MAG-11 currently includes F/A-18C squadrons, two F-35C squadrons, a fleet replacement squadron, a KC-130J tactical aerial refueling/transport squadron, and a Marine Aviation Logistics Squadron. Collectively, these units fly over 40,000 hours annually in training and operations, maintaining readiness for rapid deployment and sustained combat air operations.

The discussion ranged from the tactical versatility of the KC-130J to the strategic implications of the F-35C, revealing an organization deeply committed to realistic experimentation and learning from operational experience rather than PowerPoint presentations.

THE KC-130J: MORE THAN JUST A TANKER

Colonel DeVore's characterization of the KC-130J as the aircraft that "makes the MEF move" captures an essential truth about modern Marine aviation: flexibility is survival. During the exercise, MAG-11 was maximizing the full capability suite of the KC-130J across multiple mission sets, aerial delivered ground refueling, traditional drop zone resupply, equipment and logistics movement, personnel transport, close air support, and aerial refueling. This comprehensive employment strategy reflects a fundamental shift from platform-centric thinking to capability-based operations.

The beauty of the KC-130J platform, according to DeVore, lies not just in its multifaceted capabilities but in the adaptability of its crews. "It's not really a hard transition to say, tomorrow I need you to take this many gallons of fuel to this node, and then tomorrow I need you to go drop off folks at a DZ. And then the next day, hey, I need you to haul these certain things into a node," he explained. This operational flexibility becomes crucial in distributed operations where forces must rapidly shift mission profiles based on evolving tactical requirements.

The discussion also highlighted an often-overlooked complementarity issue within Marine aviation. With the MV-22 Osprey fleet consistently overtaxed, the potential for the KC-130J to offload some lift missions, particularly for specialized equipment like ground-based radars, represents an important force-multiplier consideration. The Osprey's unique capabilities make it a high-demand asset, and thoughtful mission allocation between platforms becomes essential for maintaining operational tempo without burning out critical assets. This raises broader questions about fleet management and the need for either additional airframes or smarter utilization of existing platforms.

I suggested that the coming of the CH-53K will help the configuration of the force and reduce some of the demands that are placed on the KC-130J and free up the Osprey for other tasks than moving supplies.

THE F-35C: BEYOND CARRIER OPERATIONS

One of the most significant aspects of the conversation centered on the Marine Corps' use of the F-35C, the carrier variant that emerged from the service's commitment to support naval aviation but has proven to be far more valuable to the Marine Corps than many initially anticipated. Colonel DeVore emphasized that beyond the obvious benefit of greater fuel capacity translating to extended time on station and increased range, the F-35C brings the same digital interoperability that makes the entire F-35 family transformative.

"One of the biggest capabilities that you're going to get with the F-35C is just going to be more fuel, which translates for more time to sense and make sense of the battle space," DeVore explained. This extended sensing capability becomes particularly significant in the distributed operations concept that the Marine Corps is developing. The longer wing of the F-35C generates more lift, enabling operations from shorter runways, a critical capability for expeditionary operations where pristine airfields may be unavailable or undesirable from a signature management perspective.

The tactical flexibility extends further. With the Marine Corps' mobile arresting gear systems, the F-35C can be stopped on runways of reduced length, and its ability to take off with less than a full fuel load and be refueled airborne creates multiple operational pathways. As DeVore noted, these characteristics allow commanders to be "smart about how you configure the aircraft" based on mission requirements and available infrastructure. This flexibility mirrors the carrier-variant's design heritage but applies it to expeditionary contexts that would have seemed improbable during the aircraft's development.

Perhaps most importantly, the F-35C serves as a critical node in the distributed kill web that the Marine Corps is building. While it lacks a dedicated Electronic Warfare (EW) suite like the EA-18G Growler, it can perform elements of the airborne surveillance and battle management functions through its sensor suite and extended time on station. The ability to see, sense, and share information across greater distances and for longer periods makes the F-35C an integral component of reach versus range. he idea that an aircraft's true operational footprint extends far beyond its physical location through networked sensors and communications.

DIGITAL INTEROPERABILITY
AND THE KILL WEB

Throughout the conversation, Colonel DeVore returned repeatedly to the theme of digital interoperability or the ability of platforms to share data seamlessly across the force. This capability represents perhaps the most significant revolution in military aviation since the jet engine, yet it remains poorly understood outside specialized communities. The F-35's ability to serve as a sensor and data node, not merely as a strike platform, fundamentally changes how commanders can employ airpower.

The integration challenge goes beyond simply connecting platforms. As the Marine Corps fields different block upgrades of aircraft, whether F-35s, F/A-18s, or MV-22s with roll-on/roll-off mission systems, maintaining awareness of what capabilities each configuration brings becomes essential. DeVore expressed confidence that the Marine Corps has solved this problem through active involvement with the Joint Program Office and rigorous configuration management. "I'm 100% confident right now that we're flying aircraft that are same page, safe, relevant and there's no confusion about what aircraft has at any given time, and then what capability set that brings," he stated.

This stands in contrast to challenges observed in other services, where aircraft in different software configurations have created confusion about capabilities at the tactical edge. The Marine Corps' approach of normalizing and demystifying new capabilities through exercises like the one at Miramar helps ensure that ground combat elements understand what the F-35 can provide, while aviation elements understand what support the ground forces require.

NORMALIZING INNOVATION
WITH GROUND FORCES

A critical observation from the discussion involved the evolution of ground combat element understanding of F-35 capabilities. Colonel DeVore acknowledged that ground forces initially showed limited enthusiasm for the platform, partly because its unique attributes weren't immediately apparent in the confined battlespaces of Afghanistan and Iraq. However, as the Marine Corps shifts focus to distributed operations across maritime domains where range, persistence, and networked sensing become paramount, ground commanders increasingly recognize the F-35's value proposition.

"This exercise is doing a great job of normalizing and demystifying these new kind of capabilities," DeVore explained. The presence of liaison officers between aviation and ground elements facilitates this mutual understanding, ensuring that both communities learn to optimize the allocation of high-end assets like the F-35 versus other available platforms. The process involves determining whether a given mission requires F-35 capabilities or whether other assets might be more appropriate, a crucial decision matrix for commanders managing limited high-demand resources.

The discussion highlighted how the F-35's reach extends far beyond its physical range through data sharing across the fleet. This networked capability means that F-35s supporting distributed operations can simultaneously monitor multiple ground elements, assess their operational status, and provide information back to command elements about force dispositions and mission progress.

THE WEAPONS CHALLENGE:
MAGAZINE DEPTH AND INNOVATION

When discussion turned to weapons systems, Colonel DeVore's concerns as a Marine aircraft group commander became apparent. "Magazine depth is something that always concerns me," he stated plainly. The challenge extends beyond simply having sufficient munitions for it encompasses having the right weapons arriving on time, weapons that remain relevant to evolving combat environments, and weapons flexible enough to be upgraded relatively quickly as threats change.

DeVore's observation about balancing hardware versus software development represents a crucial strategic question: "Should you prioritize software development that's modular, agile, that you can put on maybe something that's not a really expensive [platform]?" This question cuts to the heart of debates about weapons acquisition and employment. The lesson from ongoing conflicts, particularly the global war in Ukraine, demonstrates that sustaining forces with weapons and ammunition exceeds what democracies have proven willing to resource adequately.

The conversation explored the concept of diversifying weapons portfolios or moving beyond the assumption that every engagement requires million-dollar precision munitions. The discussion touched on potential pathways including loitering munitions, weapons from maritime autonomous systems, and more economical solutions that still achieve tactical effects. This represents a fundamental shift from platform-centric thinking to payload-centric thinking, where commanders consider what weapons are available across the battlespace rather than simply what ordnance is physically loaded on their aircraft or ships.

LEARNING THROUGH REALISTIC EXPERIMENTATION

Perhaps the most striking element of Colonel DeVore's approach involves his commitment to learning through realistic, challenging exercises rather than scripted demonstrations. The exercise at Miramar and Camp Pendleton deliberately created the hardest problem set, taking MAG headquarters Marines and requiring them to build command and control for aviation operations from the ground up in an austere location. This approach consciously avoids highly scripted, predictable exercises that validate existing procedures rather than discovering limitations and innovations.

U.S. Marine Corps Sgt. Nathan Kaufman, a geographic intelligence specialist with Marine Aircraft Group 11, 3rd Marine Aircraft Wing, monitors incoming information through a Multi-function Air Operations Center as part of Steel Knight 25 at Marine Corps Base Camp Pendleton, California, Dec. 4, 2025. The MAOC provides expeditionary aviation command and control and air surveillance capabilities essential to coordinating and sustaining Marine Air-Ground Task Force aviation operations. (U.S. Marine Corps photo by Cpl. Nikolas Mascroft).

"We want young lieutenants and sergeants and captains out here just trying things," DeVore explained, comparing the approach to interwar military experimentation. This creates a productive learning environment where some initiatives will be discarded as unworkable while others prove their value and get incorporated into training and doctrine. The iterative process identifies capability gaps and shortfalls, but more importantly, it reveals what the force can actually accomplish with organic assets and where higher echelon support becomes necessary.

This experiential approach also provides valuable feedback about when tactical commanders at the forward edge might have better situational awareness than higher headquarters, a crucial insight for distributing decision authority in contested environments where communications may be degraded. The exercise structure allows DeVore to determine both where his command needs support from higher headquarters and where his position at the tactical edge provides advantages in understanding and responding to the operational environment.

CONCLUSION: INNOVATION AS ORGANIZATIONAL CULTURE

When asked about the Marine Corps' capacity for innovation, Colonel DeVore responded with conviction: "If anybody's going to innovate, it's going to be the Marines. I think we kind of dominate that trade space right now." This confidence stems not from technological superiority but from cultural attributes, a willingness to try new approaches, accept failure as part of learning, and rapidly integrate lessons into operational practice.

The conversation with DeVore revealed an organization grappling seriously with the most difficult questions facing modern militaries: How do you sustain distributed forces in contested environments? How do you balance high-end capabilities with economic realities? How do you ensure different communities within the force understand each other's capabilities and limitations? How do you create decision frameworks that work when communications are degraded and situations are chaotic rather than controlled?

These questions don't have easy answers, and DeVore doesn't pretend otherwise. Instead, he and his Marines are working through them methodically, trying different approaches, measuring results, and building institutional knowledge about what works. This represents a fundamentally different mindset from assuming problems can be solved through better PowerPoint presentations or acquiring the next generation of technology. It acknowledges that military transformation happens through the hard work of experimentation, the honesty to recognize what doesn't work, and the discipline to incorporate lessons learned into operational practice.

As the Marine Corps continues developing its concepts for distributed operations in contested environments, leaders like Colonel DeVore provide reason for confidence, not because they have all the answers, but because they're asking the right questions and creating the conditions for their Marines to discover solutions through realistic, demanding experimentation. In an era where military challenges are growing more complex and resources remain constrained, this commitment to learning through doing may prove to be the most valuable

THE H-1 HELICOPTER'S DIGITAL REVOLUTION: HMLA-267 LEADS MARINE CORPS INTO DISTRIBUTED AVIATION OPERATIONS

HMLA-267's journey from "forgotten" attack helicopter squadron to fully digitally interoperable H-1 unit shows what it means to turn a legacy platform into a node in a distributed network.

In Steel Knight 2025, their Cobras and Hueys did not just provide close air support; they pushed and pulled data, extended the kill web, and operated from austere locations alongside the Blue Diamond.

The chapter uses their experience to illustrate how rotary-wing aviation remains central to chaos management when it gains the ability to sense, share, and survive in contested airspace.

While much attention focuses on fifth-generation fighters and unmanned systems, Marine Light Attack Helicopter Squadron 267 (HMLA-267 is quietly pioneering a transformation that may prove decisive for future combat operations.

As the first Marine Corps Light Attack Helicopter OR H-1 squadron to achieve full digital interoperability with a complete unit of employment, HMLA-267 is demonstrating how the H-1 helicopter platform, sometimes dismissed as vulnerable or obsolete, is evolving into a critical command and control node for distributed aviation operations. Their experience reveals not just technological advancement, but a fundamental reimagining of how rotary-wing aviation integrates with the joint force in contested environments.

U.S. Marines with Marine Wing Support Squadron 373, Marine Air Control Group 38, 3rd MAW and Marine Light Attack Helicopter Squadron (HMLA) 267, Marine Aircraft Group 39, 3rd MAW, disconnect a fuel hose from a UH-1Y Venom assigned to HMLA-267 at a forward arming and refueling point during exercise Steel Knight 25 on San Clemente Island, Dec. 10, 2025. U.S. Marine Corps photo by Lance Cpl. Samantha Devine.

FIRST TO FULL DIGITAL CAPABILITY

HMLA-267 holds a unique distinction within Marine aviation: they are the first squadron to field a complete unit of employment with fully upgraded, digitally interoperable, Link 16-capable aircraft. This unit consists of five AH-1Z Vipers and four UH-1Y Venoms, all equipped with the digital interoperability upgrade that fundamentally changes how these platforms operate in the battlespace.

As one squadron officer explained, "We're the first ones to really get all of them, all that hardware and everything upgraded, and then start the training syllabus, which isn't super well defined. So we're kind of the pioneer squadron." While other Marine Light Attack helicopter squadrons possess some digitally interoperable aircraft, HMLA-267 is the first to have an entire operational unit equipped and beginning to develop the tactics, techniques, and procedures for employing this capability at scale.

This pioneering status comes with challenges. The squadron is simultaneously learning the technology, developing training protocols, and proving operational concepts, all while maintaining their traditional close air support mission for the ground combat element. They are, in effect, writing the manual while flying the aircraft.

THE FORGOTTEN PLATFORM'S CRITICAL ROLE

One persistent theme in discussions about Force Design 2030 and future Marine Corps operations has been the question of rotary-wing aviation's relevance. In an era of long-range precision fires and unmanned systems, some have questioned whether helicopters, particularly attack helicopters, remain viable in contested environments.

The HMLA-267 officers pushed back firmly on this narrative. "When you say that the helicopters, the H-1 specifically, kind of were left out or forgot about, I think people just forgot that regardless of whatever fight that we're in, we are one of the ones closest to support the ground combat element, always have been, always will be," one officer stated. "We were almost always paired at the hip with the Blue Diamond."

The 1st Marine Division, headquartered at Camp Pendleton and serving as the principal ground combat element of I MEF, is commonly known as the "Blue Diamond." The name comes from its distinctive shoulder patch: a blue diamond-shaped background with a red numeral "1" and the word "Guadalcanal," commemorating the division's World War II campaign.

This relationship between HMLA-267 and the Blue Diamond isn't merely historical; it's structural. MAG-39, the parent organization for HMLA-267, is uniquely positioned as the only Marine Aircraft Group stationed on both an air station and a training base sharing Camp Pendleton with the 1st Marine Division. "We take our relationship with them and the ground combat element very, very seriously," the officer continued. "Everything we do supports them."

This proximity to the ground combat element isn't a weakness to be overcome but a strength to be exploited. While the H-1s may not possess the sensor suite of an F-35 or the standoff range of a long-range missile, they bring something equally valuable: the ability to operate in close coordination with ground forces while now possessing the digital connectivity to integrate with higher-end systems.

DIGITAL INTEROPERABILITY AS FORCE MULTIPLIER

The digital interoperability upgrade transforms the H-1 from a platform that primarily receives targeting information via voice radio to one that can participate in the broader digital network. This seemingly technical change has profound operational implications.

"Once we have that common operational picture, and we're speaking the same language, we're in the same digital ecosystem, that's where it really pays dividends," one officer explained. The upgrade enables H-1 crews to see the same tactical picture as F-35s, MQ-9 unmanned systems, and command elements and to contribute to that picture in real time.

The most immediate benefit manifests in the targeting cycle. Previously, prosecuting targets required extensive voice coordination: confirming location, verifying friendly positions, coordinating timing, and managing airspace. Each of these steps introduced delay and potential for error. "The question of where is it, or what is out there, and where is it, and then can I shoot that is something that can happen through a variety of different players," an officer noted. "Using the exquisite sensors that come off of those different systems, be they F-35, MQ-9, at least inside of the Marine Corps, and then we add assets outside of the Marine Corps to that picture as well, whether it's E-2, whether it's Growler... we can do that at a speed that previously was much more difficult to do."

This speed matters tactically and operationally. Targets can be identified by one platform, prosecuted by another, with battle damage assessment conducted by a third, all without extended voice coordination. If a target has already been engaged, H-1s can immediately shift to alternative targets without cycling back through command nets. The efficiency of prosecution, as the officers described it, is dramatically enhanced.

INTEGRATION WITH FIFTH-GENERATION SYSTEMS

A particularly significant aspect of HMLA-267's digital upgrade is its integration with F-35 operations. The F-35's sensor fusion and data-sharing capabilities are well-documented, but much of the discussion has focused on peer-to-peer F-35 networking or integration with Aegis systems. The H-1's ability to tap into this information flow represents a different dimension of fifth-generation integration.

The officers described how MV-22 Ospreys already operate with Link 16, creating a template for how rotary-wing platforms can integrate into the broader network. The H-1 upgrade builds on this foundation, but with the added dimension of being a strike platform that can both receive targeting data and contribute to the kill chain.

This integration enables what one officer called "a unified language" across diverse platforms. F-35s can detect and track targets with their sophisticated sensors, share that information via Link 16 to H-1s positioned closer to the battlefield, which can then prosecute those targets while remaining in coordination with ground forces. The alternative, voice coordination between platforms with different tactical pictures, introduces delays that can prove decisive in a fast-moving tactical situation.

AUTONOMOUS SYSTEMS INTEGRATION: THE NEXT FRONTIER

Perhaps the most forward-looking aspect of HMLA-267's digital transformation is the potential for integration with autonomous systems. This represents a paradigm shift in how rotary-wing aviation is understood and employed.

Current autonomous systems operate at "Level Two autonomy" where they can be tasked to perform specific missions but require human oversight and intervention. For maritime autonomous systems, ground-based unmanned systems, or loitering munitions, the question becomes: who tasks them, who monitors their operations, and who integrates their data into the broader tactical picture?

The digitally interoperable H-1, operating in close proximity to ground forces, represents an ideal command and control node for these systems. An H-1 crew maintaining station in support of a ground unit could task an unmanned aerial system to conduct reconnaissance, receive its sensor feed, and integrate that information with input from ground forces and higher-echelon systems, all while maintaining the ability to provide immediate kinetic support.

This concept challenges the notion that helicopters are obsolete in modern warfare. Instead, it positions them as critical nodes in a distributed network or platforms that combine mobility, persistence in proximity to ground forces, kinetic capability, and now digital connectivity.

THE REALITY OF DISTRIBUTED AVIATION OPERATIONS

The discussion of Distributed Aviation Operations (DAO) within the Marine Corps has often remained at a conceptual level, focused on dispersal to avoid adversary targeting. The HMLA-267 experience reveals both the promise and the practical constraints of distributed operations.

One officer drew on experience in the Philippines to illustrate the operational environment the squadron is preparing for. In the Indo-Pacific theater, characterized by distances, limited infrastructure, and dispersed objectives, the ability to operate from austere locations becomes crucial. The squadron has practiced expeditionary operations, demonstrating their ability to deploy, operate, and sustain H-1s from locations far from major bases.

However, the officers emphasized that realistic distributed operations require more than dispersal. "You still have to make a difference. You have to be lethal and survivable and sustainable," they noted. Sustainability becomes the limiting factor. An H-1 detachment operating from an expeditionary location requires fuel, ammunition, maintenance support, and parts. The unit must remain combat-effective, not merely dispersed.

Digital interoperability addresses part of this challenge by reducing the need for co-location to achieve coordination. H-1s need not be based with F-35s or command elements to operate in close coordination with them. They can be positioned to support ground units while maintaining digital connectivity with the broader force. This enables genuine distributed operations where distribution serves operational purpose rather than merely complicating adversary targeting.

TRAINING AND TACTICS DEVELOPMENT

As the first squadron with full digital capability, HMLA-267 faces the challenge of developing training syllabi and tactical employment concepts without established templates. This pioneering work occurs in parallel with operational commitments, creating what one officer described as learning "much more quickly" than in previous generations of technology adoption.

The squadron leverages exercises to test and refine their digital capabilities. These exercises provide opportunities to integrate with F-35s, work with ground units, and operate in joint environments. Each iteration reveals new possibilities and new challenges, informing the development of tactics and procedures.

One key learning: digital interoperability doesn't simply make existing operations easier—it enables fundamentally different approaches to mission execution. The speed of the targeting cycle, the ability to dynamically re-task based on real-time information, and the integration with diverse platforms creates opportunities that didn't exist in voice-only coordination regimes.

The squadron must also address the human factors of operating in this new environment. Crews must develop comfort with digital systems while maintaining proficiency in traditional skills. The cognitive load of monitoring multiple information streams while flying and employing weapons requires new training approaches. Building crew coordination when both pilots can access rich digital information represents a different challenge than traditional instructor-student or aircraft commander-copilot relationships.

COMMAND AND CONTROL EVOLUTION

The digital transformation HMLA-267 is pioneering has implications beyond tactical efficiency. It challenges fundamental assumptions about command and control in military operations. The squadron's experience illuminates a tension between the centralized availability of information and the decentralized execution required by operational tempo.

One officer articulated this challenge: "If commanders grow up with all these sensors providing information, and they're readily having all this information, I'm concerned about that centralized control even being more centralized... But I think the Marine Corps, if we stay true to our doctrinal execution with regards to maneuver warfare, and we go out and execute on mission-type orders... things are going to be changing in real time. We have to be able to trust the operator to execute via decentralized control and execution of mission-type orders."

This observation captures a fundamental question facing not just the Marine Corps but the joint force: does increased information availability enable more effective decentralized operations, or does it create pressure for increased centralized control? The technology enables both possibilities.

The H-1's role as a tactical-level node, operating in close proximity to ground forces, argues for the former approach. Crews can observe effects in real time, adjust to changing circumstances, and exploit fleeting opportunities. Requiring approval cycles through higher echelons would negate the speed advantages that digital interoperability provides.

Yet the temptation exists to use digital connectivity to exert greater control. Senior commanders, able to see the same tactical picture as forward units, may struggle to resist the urge to direct tactical actions. This tension isn't unique to the Marine Corps, but the Corps' doctrinal emphasis on mission command and maneuver warfare provides a framework for addressing it.

JOINT FORCE INTEGRATION

A significant aspect of HMLA-267's digital capability is its potential for joint operations. The Link 16 standard enables interoperability across services, and the officers described how this opens new opportunities for Marine H-1s to contribute to joint operations.

"As we all find ourselves working in the same digital sphere, our conversation or the language that we use becomes more unified," one officer explained. "We have the ability to contribute to the joint force and also contribute to the discussion around how to prosecute the environment and affect change in a way that we typically have not had the chance to really get an invite to the table."

This is more than symbolic. In joint operations, Marine rotary-wing aviation has sometimes operated in a supporting role, dependent on other services for situational awareness and targeting. Digital interoperability enables Marine H-1s to operate as full participants in joint kill chains, receiving targeting data from Navy E-2s or Air Force platforms and contributing their observations to the broader joint picture.

This capability becomes particularly relevant in the Indo-Pacific, where joint and combined operations will be the norm rather than the exception. Marine H-1s supporting expeditionary advance base operations or ship-to-shore operations can maintain connectivity with Navy strike groups, Air Force agile combat employment elements, and allied partners, all operating in the same digital ecosystem.

LOOKING FORWARD: THE EVOLUTIONARY PATH

HMLA-267's experience represents an early stage in a longer transformation. The squadron is proving the concept and developing initial tactics, but the full potential of digitally interoperable H-1s operating with autonomous systems in support of distributed ground operations remains to be realized.

Several evolutionary paths appear probable.

First, integration with maritime autonomous systems will become increasingly important as the Marine Corps fields unmanned surface and subsurface vessels for reconnaissance and fires. H-1s operating in support of expeditionary advance base operations could serve as C2 nodes for these systems, tasking them and integrating their data.

Second, the proliferation of small unmanned aerial systems at the tactical level will require integration into the H-1's digital picture. Rather than treating these as separate capabilities, future operations may see H-1 crews controlling or monitoring multiple unmanned systems while maintaining their traditional close air support role.

Third, improvements in data-link capacity and sensor integration will enable richer information sharing. The current Link 16 implementation represents a starting point; future upgrades will likely expand bandwidth and enable new forms of coordination.

Finally, as the Marine Corps fields the AH-1Z replacement, whatever that platform may be, the lessons learned by HMLA-267 will inform requirements and design. The next-generation attack helicopter will be designed from inception as a digital node, not retrofitted with digital capability.

CONCLUSION

HMLA-267's role as the first squadron with fully digitally interoperable H-1s positions them at the forefront of a transformation in rotary-wing aviation employment. Far from being obsolete, the H-1 is evolving into a critical node in distributed operations—a platform that combines the traditional strengths of rotary-wing aviation with the connectivity and integration capabilities of modern digital warfare.

The squadron's experience demonstrates that digital interoperability isn't merely a technical upgrade but an operational enabler that fundamentally changes how rotary-wing aviation contributes to the fight. By dramatically compressing the targeting cycle, enabling integration with fifth-generation systems and autonomous platforms, and facilitating joint operations, digitally interoperable H-1s represent a bridge between traditional close air support and future distributed operations.

As the Marine Corps continues to refine Force Design 2030 and develop concepts for operating in contested environments, HMLA-267's pioneering work provides crucial operational insights. Their ability to maintain relevance in a rapidly changing technological landscape while staying true to the Marine Corps' doctrinal emphasis on mission command and maneuver warfare suggests a path forward that leverages new capabilities without abandoning proven operational approaches.

The ultimate test will come not in exercises but in operations, where the speed, flexibility, and integration that digital interoperability enables will either prove decisive or expose limitations. But for now, HMLA-267 is writing the manual for how Marine rotary-wing aviation will operate in future conflicts, one flight, one mission, one digital upgrade at a time.

Officers in the interview:
LtCol Michael Oates, MAG-39, CO of HMLA-267
Maj Jonathan Moss, MAG-39, HMLA-267

FROM COBRA PILOT TO DIGITAL WARRIOR: THE EVOLUTION OF MARINE CORPS DIGITAL INTEROPERABILITY

U.S. Marines with Marine Light Attack Helicopter Squadron (HMLA) 267, Marine Aircraft Group 39, 3rd Marine Aircraft Wing, conduct shutdown procedures for the AH-1Z Viper and an UH-1Y Venom during Steel Knight 25 at Marine Corps Air Ground Combat Center. Photo by Lance Cpl. Parker Peichel

The transition to digital interoperability is ultimately a human transformation, and the story of one Cobra pilot's evolution into a digital warrior captures that shift. This chapter follows his path from voice-radio CAS in multiple squadrons to becoming a leading practitioner of information-age tactics, where managing data, networks, and machine-to-machine links is as important as weapons employment. It serves as the bridge between platform-centric narratives and the broader DI culture the Corps must build to make kill webs real.

Five years ago, the cockpit of a Marine Corps AH-1Z Cobra attack helicopter was a very different place. Digital interoperability was a distant concept, something discussed in Pentagon briefings but not yet realized in the ready rooms and flight lines where Marine aviators prepared for combat. Today, Major Jonathan Moss sits at the vanguard of a revolution that is fundamentally changing how Marine aviation fights, thinks, and integrates with the joint force.

During a recent interview at Marine Aircraft Group 39, Camp Pendleton, Moss shared insights into his remarkable journey from traditional attack helicopter pilot to one of the Marine Corps' leading experts in digital interoperability. His story illuminates not just the technical transformation of the H-1 helicopter community, but the profound mental shift required of warriors adapting to information-age warfare.

THE FOUNDATION: DIVERSE OPERATIONAL EXPERIENCE

Moss began flying H-1 helicopters in 2015, accumulating experience across three different fleet squadrons. He flew with the Gunfighters at Camp Pendleton, with Scarface out in Hawaii, and completed a West Pacific Marine Expeditionary Unit deployment to CENTCOM and two Unit Deployments to Japan. This broad operational background proved essential, as it exposed him to the full spectrum of Marine aviation missions, from close air support to deep strike operations.

What made his experience particularly valuable was the diversity of operating environments and tactical situations he encountered. Each squadron, each deployment, each mission set contributed pieces to what would become a comprehensive understanding of how attack helicopters integrate into the larger battle space. He learned not just how to employ weapons, but how to think spatially about coordination, deconfliction, and the synchronization of multiple aviation assets in complex objective areas.

Yet throughout these years, a critical gap existed. The information needed to prosecute targets effectively often arrived slowly, through verbal communications over VHF or UHF radio, creating delays that compressed decision-making time and limited operational effectiveness. Moss and his fellow Cobra pilots were exceptional at weapons employment and battle space management, but they operated within significant information constraints.

THE TRANSFORMATION: VMX-1 AND OPERATIONAL TESTING

The pivotal moment in Moss's transformation came when he joined Marine Operational Test and Evaluation Squadron One, known as VMX-1. This assignment placed him at the cutting edge of Marine aviation innovation, specifically working on operational testing for the H-1 digital interoperability program.

At VMX-1, Moss's role went far beyond simply learning to use new equipment. He was tasked with conducting comprehensive operational tests that examined the entire spectrum of Marine aviation functions and mission essential tasks. The challenge was to evaluate how this new capability could support existing missions and, more importantly, to explore what became possible that wasn't before.

This systematic evaluation process forced Moss to think differently about his platform and its role in the battle space. Rather than viewing digital interoperability as simply adding new hardware to the aircraft, he began to understand it as a fundamental expansion of capability that changed the domain of the possible. The question shifted from asking what tasks could be accomplished to exploring what new operational concepts could be enabled.

CROSS-COMMUNITY INTEGRATION

Perhaps most significantly, VMX-1 provided Moss with a unique opportunity for cross-community integration. He found himself working intimately with essentially an entire Marine Air-Ground Task Force worth of aviation assets. This meant sitting down across the table from F-35 pilots, MQ-9 operators, and aviators from other communities to develop integrated operational concepts. While with VMX-1, Moss held the following billets: Operational Test Director Assault Support Division, and Department Head overseeing H-1 and MV-22 operational testing.

These collaborative planning sessions proved transformative. Instead of operating within the traditional stovepipes that had long characterized military aviation, Moss learned to think in terms of integrated kill webs where different platforms contributed their unique capabilities to shared objectives. He discovered that solving complex battle space problems required bringing together experts from multiple domains, from information warfare specialists to dynamic targeting professionals, from communications SMEs to fellow aviators with different platform perspectives.

THE MENTAL REVOLUTION: CHANGING THE WARRIOR'S MINDSET

Technology alone cannot transform warfare. The most advanced digital systems are useless if the warriors who employ them remain locked in outdated mental frameworks. Moss's experience illustrates this critical truth. His journey wasn't just about learning to operate new equipment but about fundamentally changing how he thought about his role, his platform, and the nature of air combat itself.

As Moss describes it, the transformation began with recognizing that information has become as much a weapon as any missile on his wing stores. The fundamental questions driving his operations shifted from purely kinetic concerns to a more sophisticated information-centric framework: What do I know? What do I not know? How can I satisfy those information gaps? How quickly can I access the information I need to continue pursuing my objectives?

This cognitive shift was accelerated by the evolution of the threat environment itself. As Moss points out, the modern fight cannot be reduced to kinetic exchanges alone. Success in contemporary warfare requires mastering the electromagnetic spectrum. The spectrum fight, as he terms it, has become prerequisite to kinetic effects. Understanding electronic warfare, communications security, and information dominance became as essential to his professional development as mastering weapons employment.

PROFESSIONAL CURIOSITY AND INNOVATION CULTURE

Moss credits much of his successful adaptation to what he calls professional curiosity, a trait he sees pervading the attack helicopter ready room. Marine Cobra and Huey pilots, he notes, maintain an almost precocious interest in finding new ways to solve old problems. This culture of innovation, fostered through ready room discussions and collaborative problem-solving, creates an environment where aviators continuously think creatively about more efficient approaches to warfare.

Importantly, this innovation culture balances the joy of discovery with the deadly seriousness of combat. Moss describes how maintaining enthusiasm for creative problem-solving helps free the mind from the paralyzing burden of fear, even while acknowledging that self-preservation remains a fundamental concern. The goal is to work faster inside an observable battle space than adversaries can, and achieving this requires both processing information rapidly and making decisions confidently.

THE PRACTICAL IMPACT: TRANSFORMING THE TARGETING CYCLE

The practical implications of Moss's transformation become clear when examining how digital interoperability has changed the targeting cycle for attack helicopter operations. Traditionally, Cobra pilots contributed primarily at the attack phase of the find-fix-track-target-engage-assess cycle, depending on their own sensors and pre-mission briefings to locate and prosecute targets.

With digital interoperability, everything prior to the attack phase can now be assisted by other players in the battle space. Instead of depending on information delivered during pre-mission planning or transmitted via radio during flight, Moss can now access near-real-time intelligence before his aircraft even takes off. When he does launch, he knows with higher fidelity where threats are located, where other threats might harm his approach, and how the overall objective area is evolving.

The speed of this information flow represents a quantum leap. Previously, Moss might have ten to fifteen minutes of uncertainty about whether supporting F-35s had even taken off or where assault support aircraft were positioned. Now, a glance at his digital display provides immediate awareness of the common operating picture. Those extra minutes of advance knowledge allow him to think farther ahead, to plan for contingencies, to position himself more effectively for the next phase of the operation.

This capability becomes even more powerful when integrated with the exquisite sensors available on platforms like the F-35 Lightning II or the MQ-9 Reaper. These systems can detect and track targets that a Cobra's organic sensors might miss, particularly in challenging environments like triple-canopy jungle. By tapping into this network of information, Moss gains access to targeting data that dramatically enhances his effectiveness while reducing his time in vulnerable positions.

LEADING THE PACK: HMLA-267 AND THE DIGITAL FUTURE

Today, Moss serves as a department head with Marine Light Attack Helicopter Squadron 267, part of MAG-39 at Camp Pendleton. His squadron holds a unique distinction: they are the first Marine H-1 squadron to receive a complete unit of employment with digitally interoperable, Link 16 capable aircraft. This means they have all the hardware and training to operate as a fully integrated digital force, making them the pioneer squadron for this transformation across the Marine Corps.

This pioneering role brings significant responsibility. With the training syllabus still being developed, Moss and his fellow aviators are not just learning to use new systems but establishing the baseline of knowledge and best practices that subsequent squadrons will follow. They are determining what proficiency looks like, how to integrate planning processes, and how to build capability from the ground up in a community that has historically focused exclusively on kinetic excellence.

Moss's goal is to ensure that every pilot checking into the squadron develops familiarity with tactical data links and digital interoperability from day one, in much the same way that fixed-wing aviators undergo deep syllabi on these topics. He wants the next generation of attack helicopter pilots to discuss digital interoperability as fluently as they discuss close air support procedures, making it a fundamental rather than an advanced skill.

CONCLUSION: THE HUMAN ELEMENT IN MILITARY TRANSFORMATION

Major Jonathan Moss's journey from traditional Cobra pilot to digital interoperability expert illuminates a profound truth about military transformation: technology alone changes nothing. The most sophisticated systems remain inert without warriors willing and able to reimagine how they fight. Moss's success came not from the equipment upgrade to his aircraft, but from his willingness to undertake an intellectual upgrade to his own mental frameworks.

His path offers lessons for the broader defense community grappling with rapid technological change. Diverse operational experience provides the foundation for understanding context and complexity. Formal testing and evaluation roles create opportunities for systematic exploration of new capabilities. Cross-community collaboration breaks down stovepipes and enables integrated thinking. Professional curiosity drives continuous innovation. And ultimately, changing the domain of the possible requires changing how warriors think about the possible.

As HMLA-267 leads the Marine Corps into the digital future, they carry forward not just new technology but a new way of thinking about attack helicopter operations. In an era where information dominance increasingly determines tactical success, Moss and his fellow pilots represent the vanguard of a transformation that extends far beyond their platform, pointing toward a future where every Marine aviator must master both the kinetic and the digital dimensions of warfare.

FROM KILL CHAINS TO KILL WEBS: 3RD MARINE AIRCRAFT WING'S COMMUNICATIONS REVOLUTION

Staff Sgt. Jonathan Negron-Cancel and Senior Airman Jacob Schoenwetter-Nolan, 920th Communications Flight radio frequency operations technician, review equipment after assembling the line-of-sight mast system during Exercise Steel Knight 25 at Los Alamitos Army Airfield, California, Dec. 4, 2025. U.S. Air Force photo by Master Sgt. Darius Sostre-Miroir.

At 3rd Marine Aircraft Wing headquarters, a group of communications and intelligence professionals are working through one of the most fundamental challenges facing modern military operations: how to separate the sensor from the shooter while maintaining the speed and precision required for contested combat. This isn't an abstract doctrinal exercise. It's the practical reality of what future Pacific operations will demand.

Within 3rd Marine Aircraft Wing, the G-6 is the commanding general's staff section responsible for communications and information technology: it plans, builds, secures, and manages the command-and-control (C2) and IT networks that let the wing command, coordinate, and fight. The G-6 provides a number of capabilities and functions for 3rd MAW:

- Provides the 3rd MAW commander with planning, management, and execution of C2 systems so the wing can exercise tactical and operational control of aviation forces in garrison and deployed.
- Serves as the principal advisor on communications, IT, and cyberspace matters, integrating these into operations and exercises across the wing.
- Plans, installs, operates, and maintains the wing's communications and IT infrastructure, including classified and unclassified networks, telephony, data services, and collaboration tools.
- Designs and manages the communications architecture for exercises and operations so aviation C2, fires, logistics, and support elements can share data and orders reliably..
- Enforces cybersecurity and information assurance policies on wing networks, protecting data, systems, and services against intrusion and disruption.
- Oversees COMSEC (communications security) management to ensure voice and data links used by 3rd MAW remain secure and compliant with higher headquarters directives.
- Coordinates with higher and adjacent commands (e.g., I MEF, MCEN entities, and joint/coalition partners) to ensure interoperability of networks and C2 systems during joint and combined operations.
- Supports information environment and electromagnetic

spectrum operations by aligning network and data service priorities with the intel, fires, and IE Ops planners.

- Advises on IT resourcing, modernization, and lifecycle management of communications and C2 capabilities used by 3rd MAW headquarters and subordinate units.
- Develops and enforces wing-level C4/IT standards, contributing to broader Marine Corps enterprise network initiatives and ensuring readiness of communications capabilities.

The conversation with the G-6 team reveals something often overlooked in discussions about military transformation: the revolution isn't just about new platforms or weapons systems. It's about fundamentally reimagining how information flows through combat operations, how commanders make decisions under pressure, and how forces separated by vast distances can coordinate effects with precision and economy.

BEYOND THE KILL CHAIN

"We're trying to design a kill chain that separates the sensor from the shooter," explained one of the G-6 officers, cutting straight to the operational challenge. "Both adversary and friendly munitions now go further than our sensors can see from the shooter itself. To achieve true standoff, you have to leverage sensors that aren't on board the shooter."

This seemingly simple observation carries profound implications. The traditional model of a platform carrying both its sensors and weapons worked well enough when engagement ranges were relatively short. But modern missiles and precision weapons have outpaced the sensor range of individual platforms. An air platform might be able to launch a weapon 100 miles away, but can it see the target at that range? Often not. This creates what military planners call an "engagement zone" problem.

The solution requires creating a network of sensors positioned to give shooters the data they need, when they need it, with the precision modern weapons demand. "We need to create an engagement zone of sensors to allow the shooters to actually do the kills that we need to do," as the discussion framed it.

This isn't just a technical challenge. It's an organizational one, requiring Marine aviation, ground forces, naval elements, and potentially joint partners to share information with unprecedented speed and reliability. The digital link between sensor and shooter becomes the critical enabler, and that's where the communications professionals come in.

THE COMMUNICATIONS ARCHITECTURE BEHIND MODERN OPERATIONS

For the G-6 team at 3rd Marine Air Wing, their mission is clear: ensure that critical targeting data can flow from sensor platforms to shooters regardless of domain, service, or location. This means leveraging everything from space-based sensors to terrestrial ISR platforms, from Link 16 tactical data networks to commercial satellite communications.

"We're trying to leverage a complement of terrestrial sensors and space-based sensors," one officer explained, describing the find-fix-track portion of the kill chain. "Then when it gets to target and engage, we're really trying to leverage Link 16 as our primary path to get that message to an airborne shooter."

But here's where it gets complicated. Link 16 is a proven tactical data link, but it's not the only communications path available.-The G-6's approach reflects a practical reality: redundancy is survival.

"I have a personal rule of thumb," said one officer: "I need three ways to win, three ways of getting a message across the wire. I'll take commercial SATCOM, whether it's Starlink or ViaSat. I'll take cellular. I'll take fiber. I'll take anything."

This comprehensive approach to communications reflects a fundamental shift in military thinking. The old model assumed that military-specific, highly secure communications would always be available and sufficient. The new reality acknowledges that in contested environments, communications will be disrupted, jammed, or denied. Having multiple pathways, including commercial systems, military SATCOM like the V-SAT family, and even leased DISA circuits, provides the resilience that operations require.

SPEED, ADAPTATION, AND
THE UKRAINIAN EXAMPLE

The conversation repeatedly returned to a theme that has dominated military thinking since 2022: the lessons of Ukraine. But not in the way many analysts discuss the conflict.

"The Ukrainians' ability to adapt is key, but it has to include speed," an officer emphasized. "The speed of adaptation is critical. Going back to Commandant Gray's warfighting doctrine and maneuver warfare, it's about speed at which a commander can make decisions. AI won't make us win by itself, but it can prioritize what's important to help us key in on what's essential in our thinking process."

This gets at something deeper than just buying new technology. The Ukrainian forces have demonstrated an ability to rapidly integrate commercial drones, modify tactics, and create new operational approaches faster than their adversary can adapt. In one operation, when Russian forces penetrated several miles into Ukrainian territory using motorcycles and tanks, Ukrainian defenders used a combination of HIMARS rockets, APCs, and drones in a coordinated strike they had never practiced before. That's the kind of speed and flexibility that future operations will demand.

But speed requires more than just willingness to experiment. It requires empowering junior leaders to make decisions based on commander's intent rather than waiting for specific authorization. "How do we empower our young field grade officers and staff NCOs to execute on commander's intent?" an officer asked. "There's a sergeant who understands what's truly going on, speeding information up the pipe. How do we change that so the sergeant in the battlefield has full awareness?"

This connects directly to the kill web concept. If every sensor can potentially become a shooter, or enable a shooter, then the traditional hierarchical command structure becomes too slow. The decision to engage a target can't always wait for authorization to travel up the chain of command and back down. There has to be a balance between fires control authority and operational necessity.

THE JOINT SYNCHRONIZATION CHALLENGE

One of the major challenges is the question of targeting authority and joint synchronization. When operating across vast Pacific distances with limited resources, how do you ensure that Air Force, Navy, and Marine Corps elements aren't redundantly engaging the same targets while missing others?

"In the modern battlespace, especially across the Indo-Pacific, we need to synchronize joint effects across all domains in a very timely and well-coordinated manner," explained one of the officers. "Some of these higher tier threats require synchronized effects that operational commanders need to plan carefully."

This creates a tension between tactical flexibility and strategic coordination. At the tactical level, speed matters. A fleeting target might only be vulnerable for minutes. But at the strategic level, striking certain targets might have implications for escalation control, coalition politics, or nuclear signaling.

COMMERCIAL INTEGRATION AND UNCLASSIFIED NETWORKS

One of the most interesting developments discussed was the push to integrate sensors through unclassified networks wherever possible. This reduces the security management footprint, eliminates the need for extensive encryption overhead, and allows for much more rapid experimentation and integration.

"We've pushed out a lot of these sensors, and the data paths come back and enter a central point, fusing the sensor inputs," explained one officer. "A lot of that we've been experimenting with riding over unclassified networks to make it more agile and expand the pool of usable data."

This has profound implications. During the Steel Knight exercise,

3rd MAW has been working to integrate unmanned water vessels operated by 1st Marine Division into the common operational picture. The goal is to take sensor data from these platforms and proliferate it throughout the network so that any shooter can potentially use it.

"Last year we had UAVs operating and the adjacent units couldn't see them in their system," an officer noted. "They could see them with their eyeballs, but not in their picture. That's a problem. Now we're working on including that in the air picture, because eventually those systems become targets that any sensor should be able to identify and any shooter should be able to engage."

This gets at the heart of the kill web concept: any sensor should be able to support any shooter. The platform doesn't matter. What matters is the effect you're trying to create and having the communications architecture to enable it.

THE GENERATIONAL ADVANTAGE

An interesting sideline in the discussion touched on generational differences in approaching technology. The younger Marines and sailors coming into the service have grown up in a digitally connected world. They're more comfortable experimenting with new systems, more willing to take risks with unfamiliar technology.

"The younger folks are more willing to incorporate new technologies," an officer observed. "They'll play with a gadget and figure it out. They can look at a new device and immediately figure out how to log on, take a picture, make it work. That willingness to experiment is valuable."

But this cultural advantage only matters if leadership creates an environment where that experimentation is encouraged and failure is treated as a learning opportunity rather than a career-ending mistake. The Marine Corps' emphasis on mission command and commander's intent theoretically enables this, but organizational culture doesn't always match doctrinal aspiration.

THE FORCE DESIGN DILEMMA

The conversation touched on the broader challenges facing the Marine Corps as it navigates the transition to Force Design 2030 (now just "Force Design" without a specific timeline). The service has made deliberate trades, accepting reduced readiness in some areas to invest in future capabilities.

"We're in a crux period where we need forces ready now while we're still building for the future," an officer acknowledged. "The infantry has challenges with force size. We're transitioning from fourth-generation to fifth-generation fighter attack aircraft. We're waiting on TPS-80 radar systems that were delayed by Air Force procurement decisions."

This creates a fundamental tension. The Marine Corps has always prided itself on being adaptable, on being the force-in-readiness that can respond to crises. But that readiness is based on having mature, proven systems and well-rehearsed tactics. Transformation means accepting periods of reduced capability while new systems are integrated and new tactics are developed.

The G-6's role in this transformation is crucial but often invisible. While attention focuses on new platforms, F-35s, CH-53Ks, unmanned systems, the communications architecture that allows these platforms to work together is what actually enables new operational concepts.

THE LOGISTICS
ELEPHANT IN THE ROOM

Perhaps the most sobering part of the discussion centered on logistics. All the advanced communications in the world don't matter if forces can't be sustained in theater. "Contested logistics" is the euphemistic phrase, but the reality is starker: the U.S. military currently lacks the capability to sustain distributed forces across Pacific distances against a peer adversary.

Modern logistics depends heavily on IT systems. This points to a broader truth: the communications ecosystem that enables kill webs also have to enable logistics webs, intelligence webs, and all the other functions that military operations require. It's not just about putting warheads on foreheads. It's about creating an integrated ecosystem where information flows efficiently enough to enable economy of force across all functions.

LOOKING AHEAD: STEEL KNIGHT AND BEYOND

Steel Knight exercises provide the Marine Corps with opportunities to rehearse these complex operations in conditions short of actual combat. The value isn't just in testing equipment, though that matters. It's in building the muscle memory for how units communicate, coordinate, and make decisions under pressure.

The G-6 team's work in these exercises is foundational. They're building the digital infrastructure that allows a distributed force to act as an integrated whole. They're experimenting with commercial systems, testing new data paths, and working through the hard organizational questions about who needs what information when.

But exercises also reveal gaps. Every Steel Knight exposes shortfalls in equipment, doctrine, or training. The question is whether the acquisition system and organizational culture can adapt rapidly enough to address those gaps. The Ukrainian example suggests that rapid adaptation is possible when necessity drives it and bureaucracy doesn't block it.

The challenge for the Marine Corps and for the joint force more broadly is whether it can develop that culture of rapid adaptation in peacetime.

- Can we create organizations agile enough to integrate new capabilities in months rather than years?
- Can we empower junior leaders to experiment and fail without career consequences?
- Can we build the communications infrastructure that modern operations demand?

CONCLUSION: THE ECOSYSTEM MATTERS

What the conversation with 3rd MAW's G-6 reveals is that modern military operations rest on an ecosystem of communications, logistics, and organizational culture that has to work seamlessly. The platforms get the attention, the F-35s, the unmanned systems, the long-range missiles. But without the communications architecture to coordinate them, they're just expensive individual platforms.

The kill web isn't just a technical concept. It's an organizational one. It requires separating sensors from shooters but maintaining the links between them. It requires empowering junior leaders while maintaining strategic coherence. It requires integrating joint and coalition partners while protecting essential secrets. It requires being agile enough to adapt rapidly while maintaining the discipline and professionalism that military operations demand.

The G-6 professionals working these challenges don't have all the answers. But they're asking the right questions, experimenting with new approaches, and building the foundations for how future forces will operate. In an era of major power competition, that might be the most important work happening anywhere in the Department of War.

Participating in the conversation were the following Marines:

- 1stlt Richard Patton
- LtCol Richard Larger
- MGySgt Adam White
- Maj Joseph Laffey
- Maj Brendan Mulcahy
- Col Michael Anthony

MACG-38: THE COMMAND-AND-CONTROL ENABLER FOR MARINE CORPS DISTRIBUTED OPERATIONS

U.S. Marine Corps Cpl. Ezekiel Rouse, a tactical air defense controller with Marine Air Support Squadron 3, Marine Air Control Group 38, 3rd Marine Aircraft Wing, monitors aircraft locations from the Multifunction Air Operations Center in support of Exercise Steel Knight 23.2 at Marine Corps Base Camp Pendleton, California, Dec. 6, 2023. U.S. Marine Corps photo by Cpl. Daniel Childs.

Behind every distributed aviation vignette in Steel Knight sits MACG-38, quietly wiring together sensors, shooters, and decision-makers. This chapter examines how the group is redesigning Marine air command and control for a world of expeditionary spokes, FARPs, and stand-in forces, rather than fixed towers and big-wing radars parked at secure bases. It shows MACG-38 as both a technical integrator and a doctrinal pioneer for chaos-management C2.

The modern Marine Air Control Group 38 (MACG-38) represents a fundamental shift in how the Marine Corps conceptualizes aviation capabilities in the era of distributed operations.

Rather than viewing Marine aviation through the traditional lens of individual platform types, MACG-38 demonstrates how integrated command, control, communications, and air defense functions serve as the essential enabler for the entire aviation enterprise.

Colonel O'Connell IV, commanding officer of MACG-38, articulates a vision where the control group functions as the dial for force configuration, dynamically assembling tailored packages to create specific impacts across time and space.

THE CONTROL GROUP
AS INTEGRATION HUB

MACG-38's organizational structure reveals its central role in Marine aviation operations. The control group encompasses all operational support for the air wing beyond the flight line itself.

This includes the Marine Wing Support Squadrons (MWSS) handling ground support, engineering, fuel, firefighting, and the physical infrastructure needed to generate sorties. The communications squadron provides the kill web design backbone or the sensor and communication networks that enable distributed operations.

At Camp Pendleton, the 3D Low Altitude Air Defense (LAAD) Battalion brings counter-UAS and integrated air defense capabilities, while the Marine Air Support Squadron (MASS-3) houses the tactical air controllers who provide real-time aircraft control and airspace deconfliction. Marine Air Control Squadron (MACS) 1 provides fixed site and expeditionary air traffic control services and also provides the expertise and infrastructure for the 3d MAW Tactical Air Command Center (TACC).

While the MACG has been in existence for decades, there is a newfound appreciation for the criticality of the unit, its Marines, and the capabilities they bring to bear in future warfare scenarios.

This integration represents a fundamental evolution in how Marine aviation capabilities are conceived and employed. Rather than discrete units operating independently, MACG-38 demonstrates platform-agnostic thinking where the focus shifts from specific aircraft to the effects and impacts that can be achieved through properly integrated force packages.

The control group serves as the nexus where sensors, shooters, and command elements connect to create what Colonel O'Connell describes as the kill web or a networked approach to operations that moves beyond linear kill chains.

FROM CRISIS MANAGEMENT
TO CHAOS MANAGEMENT

In my view, the operational concept driving MACG-38's evolution reflects a broader shift in how military forces must operate in contemporary conflict environments whereby the notion of chaos management rather than traditional crisis management is central. This distinction is crucial: crisis management assumes a knowable problem set that can be resolved to restore stability, while chaos management recognizes that modern operational environments are characterized by persistent complexity and ambiguity.

In chaos management, forces must probe the environment through action to understand what they're actually confronting. The Marine Corps, particularly through units like MACG-38, is uniquely positioned to serve this exploratory function. Unlike larger Air Force or Navy formations that signal major commitment through their very presence, Marine force packages can be calibrated to create specific impacts while maintaining proportionality. This measured approach allows commanders to test adversary responses, gather intelligence about intentions and capabilities, and shape the operational environment without necessarily triggering escalation.

The control group's ability to task-organize enables this flexibility. When given a mission, MACG-38 doesn't deploy as a fixed unit but rather assembles the precise slice of capabilities required. For a Marine Expeditionary Unit deployment, this might include an LAAD detachment for air defense, air controllers for aviation coordination, MWSS elements for engineering and sustainment, and communications squadron personnel for the digital backbone. For other missions, entirely different combinations might be required. This modularity transforms the control group from a static organization into a dynamic force generator.

THE AIR DEFENSE RENAISSANCE

One of the most significant developments Colonel O'Connell describes is the renaissance of Marine Corps air defense capabilities. For decades, LAAD battalions existed on the periphery of Marine Corps operations. After divesting the Hawk missile system in the late 1990s and transitioning to a Stinger-only low altitude air defense posture, these units became primarily focused on point security for maneuver forces. During Iraq and Afghanistan, LAAD battalions largely functioned as provisional infantry, as the air threat environment didn't require dedicated air defense forces.

The proliferation of unmanned aerial systems has fundamentally changed this calculus. The drone threat, from small quadcopters to sophisticated autonomous systems, has elevated air defense from peripheral to central in Marine Corps planning. Force Design initiatives have recognized this shift, growing the LAAD battalion structure significantly. The recent standup of 1st LAAD Battalion in Hawaii, designated as an experimental unit, signals the service's commitment to rapid innovation in counter-UAS capabilities.

This growth isn't simply about fielding more systems. Colonel O'Connell emphasizes that the LAAD community is being asked to think creatively about the tools needed to address the drone challenge without bankrupting the force. The economics of air defense have become critical, using million-dollar missiles against thousand-dollar drones represents an unsustainable approach. Instead, the focus has shifted toward a weapons room concept: a suite of capabilities spanning electronic defeat, kinetic options at various ranges, and integrated systems that can be tailored to specific threat profiles.

The target-to-weapon matching problem exemplifies the broader challenge of distributed operations. Rather than one-size-fits-all solutions, commanders need the ability to dial appropriate weapons against specific threats. This requires not just technology but also the operational concepts and command structures to employ these capabilities effectively. MACG-38's role in integrating ground-based air defense with airborne sensors and shooters makes it central to solving this problem.

THE F-35 INTEGRATION REVOLUTION

The F-35's introduction represents a watershed moment for Marine aviation, though its full implications are still being absorbed. Colonel O'Connell describes the aircraft's capabilities in terms of expansion, expanding the kill web, extending reach, and fundamentally changing what's possible in distributed operations.

Beyond its kinetic capabilities, the F-35 functions as a sensor and information node within the broader kill web. Its ability to collect, process, and distribute information in real time transforms the operational picture for ground forces and other aircraft. The control group's role in receiving and disseminating this information becomes crucial. While direct datalinks between F-35s and individual ground units represent one mode of operation, the more flexible approach involves F-35s feeding information to MACG-38's C2 nodes, which can then distribute relevant information through various communications pathways to whoever needs it.

This information architecture reflects platform-agnostic thinking. The value isn't in any single platform but in the network of capabilities that platforms enable when properly integrated. The F-35 can warn forces of threats, provide targeting information, or confirm that friendly forces should reposition, all without necessarily employing weapons itself. This multi-mission flexibility, combined with the control group's ability to integrate and disseminate information, creates options that didn't exist in previous generations of Marine aviation.

THE PAYLOAD REVOLUTION AND MARITIME AUTONOMOUS SYSTEMS

Colonel O'Connell's discussion touches on what I have called the payload revolution or the recognition that capabilities increasingly reside in sensors, communications systems, and effectors that can be distributed across multiple platforms rather than being tied to specific airframes. This thinking has profound implications for how Marine aviation conceptualizes force structure and employment.

Maritime autonomous systems offer particularly promising opportunities. Unlike aerial systems that face significant challenges with autonomy levels and endurance, maritime platforms operating on the surface can carry useful payloads for extended periods. The Navy's experimentation with unmanned surface vessels carrying various sensor and communications packages demonstrates the concept. For MACG-38, the question becomes: what payloads positioned where would enhance our ability to command, control, and employ aviation assets?

This could include communications relays to extend reach, sensors to provide early warning or targeting information, or even kinetic effects. The key insight is that these capabilities don't need to be on manned platforms or traditional military systems. Commercial off-the-shelf solutions, rapidly prototyped systems, and experimental platforms can all contribute to the kill web if properly integrated. The control group's role becomes less about controlling specific platforms and more about orchestrating a network of capabilities to achieve desired effects.

The challenge lies in acquisition and experimentation processes. Colonel O'Connell describes 1st LAAD Battalion's designation as an experimental unit, working directly with the Marine Corps Warfighting Lab and industry to rapidly prototype and test capabilities outside traditional acquisition channels. This model needs to expand across the control group's mission areas. The goal isn't just to experiment but to create pathways for successful concepts to transition into programs of record that can be fielded at scale.

INTEGRATING AIR DEFENSE INTO THE FIRES PARADIGM

One of the most significant conceptual shifts Colonel O'Connell describes is the integration of air defense fires into the broader fires paradigm. Historically, air defense operated somewhat independently, providing point security or area defense but not integrated into schemes of maneuver and fires in the way artillery or close air support would be.

The contemporary operating environment demands different thinking. When ground forces maneuver, they must consider the air threat, not just from traditional aircraft but from a spectrum of unmanned systems. When planning fires, commanders must account for how air defense contributes to protecting shooters, enabling maneuver, and creating windows of opportunity. The merger of the Direct Air Support Center (DASC) and Tactical Air Operations Center (TAOC) functions into a common controller reflects this integration.

The Common Aviation Command and Control System (CAC2S) provides the technical backbone for this integration, while the G/ATOR radar adds sensor capability that feeds the kill web. MASS-3's tactical air controllers provide the human element, real-time control and deconfliction that allows complex air operations in contested environments. Together, these elements enable what Colonel O'Connell describes as an integrated air and missile defense system where Marine Corps capabilities mesh with longer-range Army and Navy systems to provide defense in depth.

The weapons-to-target matching problem becomes central in this integrated approach. Against peer adversaries with sophisticated air forces and diverse unmanned systems, Marine forces need layered defenses spanning counter-UAS systems at close range, medium-range systems for various threat profiles, and integration with higher-echelon systems for extended-range threats. The weapons room concept, selecting appropriate tools for specific threats, requires both the physical capabilities and the command structure to employ them effectively. MACG-38 provides that command structure.

FORCE PACKAGING FOR DISTRIBUTED MARITIME OPERATIONS

Distributed Maritime Operations represent the operational concept driving much contemporary naval thinking, and Marine aviation must adapt to enable this approach. The control group's ability to task-organize becomes essential in the maritime environment where geography, distances, and the nature of sea control create unique challenges.

Consider Expeditionary Advanced Base Operations_(EABO) focused on fires or sensing missions. The control group provides tailored support, LAAD elements for air defense, air controllers for aviation coordination, MWSS engineers and support personnel to sustain operations, and communications packages to maintain connectivity with the broader force. The specific mix depends on the mission: a sensing-focused AEB might emphasize communications and air control, while a fires-focused location might prioritize air defense and sustainment for expeditionary operations.

The challenge lies in balancing survivability with lethality across time. Every force package exists on a survivability-lethality curve. At some point, the force becomes more vulnerable than effective. A particular capability package might create significant effects for 72 hours but becomes increasingly vulnerable beyond that window. Planning must acknowledge these limitations and build in either rotation, reinforcement, or extraction accordingly.

This realism about temporal constraints distinguishes serious operational planning from briefing chart fantasies. Commanders need forces they will actually employ, not concepts that look good on PowerPoint but create impossible dilemmas in execution. The control group's value lies partly in providing realistic assessment of what force packages can accomplish within what timeframes, enabling commanders to match capabilities to requirements with appropriate confidence.

THE COMMUNICATION
SQUADRON INNOVATION ENGINE

The communications squadron within MACG-38 represents a hub of innovation for kill web development. Co-located with a battle lab and working closely with the Marine Corps Test Support Activity at Camp Pendleton on Project DYNAMIS, the squadron experiments with emerging communications and networking capabilities. This isn't just about technical specifications: it's about understanding how different communications pathways enable or constrain operations.

In distributed operations, communications architecture becomes as important as any weapons system. How do dispersed elements maintain connectivity? What redundancy is required? How can commercial solutions supplement military networks? What protocols enable rapid formation of ad hoc networks as force packages assemble and disperse? These questions drive experimentation that feeds directly into operational concepts.

The squadron's role extends beyond providing communications services to actively designing the digital backbone that enables kill web operations. When planning an operation, communications aren't an afterthought but central to concept development. This requires communications personnel to think operationally about effects rather than technically about systems, a cultural shift that MACG-38 is actively fostering.

PLATFORM AGNOSTIC THINKING
FOR THE AUTONOMOUS ERA

Perhaps the most significant theme in Colonel O'Connell's discussion is the shift toward platform-agnostic thinking. Rather than asking what platforms the Marine Corps needs, the question becomes what capabilities are required and how can they be most effectively delivered. This inverts traditional acquisition and force development logic.

Sensors provide a clear example. Rather than integrating specific sensors onto specific platforms through lengthy acquisition programs, platform-agnostic thinking asks: what sensing capability is needed, where does it need to be positioned, and what platforms (manned, unmanned, or simply emplaced) can carry it most effectively? The answer might be multiple platforms carrying the same sensor, or different sensors optimized for specific platforms, or even expendable sensor packages that can be rapidly emplaced and replaced.

This thinking requires comfort with expendability. Not every capability needs to be on an expensive, survivable platform. Some sensors, communications nodes, or even effectors might be deliberately expendable if they provide the capability needed at acceptable cost. The key becomes the network architecture that allows individual nodes to be lost without degrading overall capability, resilience through redundancy rather than hardening of individual platforms.

For MACG-38, platform-agnostic thinking means focusing on integration rather than platforms. The control group doesn't care whether a sensor is on an F-35, an unmanned aerial vehicle, a maritime autonomous system, or an emplaced ground sensor. It cares that the sensor data flows into the kill web where it can be exploited. Similarly, communications pathways matter more than specific communications platforms. This abstraction of capability from platform enables flexibility impossible when thinking in terms of specific systems.

STEEL KNIGHT AND OPERATIONAL EXPERIMENTATION

The Steel Knight exercise series exemplifies how MACG-38 approaches operational experimentation. Rather than single-focus force-on-force exercises, Steel Knight incorporates multiple vignettes testing different aspects of distributed operations simultaneously. This reflects the reality that future conflicts won't resemble clean force-on-force engagements but rather complex environments where multiple challenges exist simultaneously.

Steel Knight allows testing of force packaging concepts: what combinations of capabilities can achieve what effects? It provides opportunity to stress communications networks and command relationships under realistic conditions. Air defense integration, aviation coordination, sustainment concepts, and command and control architectures all get exercised together rather than in isolation. This systems-level approach reveals dependencies and friction points that single-system tests would miss.

The exercise also serves as venue for introducing experimental capabilities and concepts. This creates a feedback loop between innovation and operational employment: experiments inform operational concepts, which drive requirements for new experiments.

RETHINKING THE MARINE AVIATION NARRATIVE

Colonel O'Connell's insights point toward a fundamental rethinking of how Marine Corps aviation capabilities are presented and understood. The traditional narrative focuses on type-model-series aircraft, F-35s, MV-22 Ospreys, AH-1Z attack helicopters, and so forth. While these platforms matter, they don't capture what makes Marine aviation distinctive.

The alternative narrative positions MACG-38 and the control group at the center, with type-model-series aircraft as tools that can be orchestrated to achieve effects. This reorientation makes clear that Marine aviation isn't about individual platforms but about integrated capabilities. The value proposition becomes: "We can assemble force packages tailored to specific requirements, sustain them for defined periods, and create impacts across multiple domains through integrated command, control, sensors, and fires."

This narrative better explains why Marine aviation differs from Air Force or Navy aviation. It's not primarily about the hardware, other services operate some of the same platforms. The distinction lies in the integration approach, the flexibility of force packaging, and the focus on enabling maneuver forces operating in contested environments. MACG-38 embodies this integration, making it central to the Marine aviation story rather than a supporting element.

CONCLUSION: THE C2 ENABLER FOR DISTRIBUTED OPERATIONS

MACG-38's evolution illustrates how the Marine Corps is adapting to the demands of distributed operations in an era of great power competition and persistent chaos. Rather than preparing primarily for major conventional war, the focus shifts to building capabilities that create impacts across the spectrum of competition from presence operations through high-intensity combat.

The control group serves as the essential enabler for this flexibility. Its ability to task-organize force packages, integrate diverse capabilities into kill webs, provide resilient command and control, and sustain expeditionary operations makes it indispensable for distributed maritime operations. The growth of air defense capabilities, integration of emerging technologies, and platform-agnostic thinking position MACG-38 at the forefront of Marine Corps transformation.

Colonel O'Connell's vision emphasizes realism about capabilities and constraints, honest assessment of temporal limitations, and creative thinking about how to leverage emerging technologies. This pragmatic approach, combined with aggressive experimentation through exercises like Steel Knight and experimental units , positions Marine aviation to remain relevant in rapidly evolving operational environments.

The challenge ahead lies in translating these operational concepts into acquisition decisions, training programs, and force structure. The transformation isn't primarily about buying new platforms, rather it's about building the networks, developing the doctrine, and creating the organizational culture that enables truly distributed operations. MACG-38 demonstrates that this transformation is already underway, providing a model for how command and control capabilities enable rather than constrain operational flexibility in the autonomous era.

CHAPTER 10

MARINE C2'S SHIFT FROM ENDURING BASES TO EXPEDITIONARY HUBS

The shift from big, enduring C2 bases to agile expeditionary hubs is one of the most important architectural changes Steel Knight 2025 is stress-testing. Here I describe how Marines experimented with relocating C2 nodes, managing bandwidth and risk, and sustaining decision-quality information without the comfort of fixed infrastructure. The chapter connects this basing transition to the larger Indo-Pacific problem of surviving long-range precision fires while still orchestrating a coherent fight.

Lieutenant Colonel Nicholas Astacio commands Marine Air Control Squadron One (MACS-1), a unit at the forefront of one of the most significant transformations in Marine Corps aviation command and control. His unit is part of the 3rd MAW's Marine Air Control Group 38.

During my recent visit to MCAS Miramar for the Steel Knight 2025 exercise I had a chance to talk with him about the evolving capabilities and approach of distributed C2 in Marine Corps evolution.

With 28 years in the Marine Corps, including 13 years specializing in aviation command and control, Astacio offers a unique perspective on how technological innovation, operational necessity, and strategic vision are reshaping how Marines detect, track, and engage threats in an era of peer competition.

His journey through the evolution of Marine air command and control capabilities provides a window into both the remarkable progress achieved and the challenges that remain as the Corps works to implement its Force Design 2030 vision in an increasingly contested operating environment.

Astacio's command and control career began in 2011 at an air traffic control company in Beaufort, South Carolina, where he immediately deployed to Iraq's Al Taqaddum Air Base. The equipment he worked with then represented what he now calls "big box equipment" or large, mobile radar systems with enormous footprints designed for enduring presence rather than expeditionary maneuver.

"We had large radars, two big radars, one that provided precision approach, one that provided airport surveillance radar," Astacio recalls. The system included separate command and control systems, all requiring substantial lift capability and extended setup times. "It really wasn't mobile from point A to point B. You set it up and it's going to be enduring."

This equipment paradigm made sense in the permissive environments of Iraq and Afghanistan, where the primary threats were indirect fire and mortars rather than sophisticated air defense systems or long-range precision strikes. Marines weren't worried about peer adversaries shutting down their command-and-control nodes with electronic warfare or kinetic strikes. Mobility and signature management were secondary concerns to providing reliable air traffic control and airspace management.

The strategic shift toward major power competition changed everything. As potential adversaries developed sophisticated long-range strike capabilities, anti-access/area denial strategies, and advanced sensors, the Marine Corps recognized that large, static command and control nodes would become high-value targets in contested environments. The Corps needed command and control capabilities that could move, survive, and operate within the weapon engagement zones of peer competitors.

U.S. Marine Corps Sgt. Maj. Alexander Caro (right) incoming sergeant major, receives the sword of office from Lt. Col. Nicholas Astacio, left, Commanding Officer, Marine Air Control Squadron 1, during a relief and appointment ceremony on Marine Corps Air Station Yuma, Arizona, Aug. 8, 2025. Photo by Lance Cpl. Elizabeth Gallagher (Marine Corps Air Station Yuma).

Astacio's tour in Okinawa with the Marine Tactical Air Command Squadron illustrated another phase of evolution. The unit provided command post functionality for First Marine Aircraft Wing but reflected an organizational structure that Force Design 2030 would fundamentally reshape.

"That unit doesn't exist anymore," Astacio explains. The transformation wasn't simply about renaming units but about reimagining how command and control capabilities integrate with distributed operations. The squadron was reorganized into a company-level element within the Marine Air Control Squadron structure, part of a broader effort to create more agile, deployable command and control packages aligned with the emerging Littoral Anti-Air Battalion concept.

The goal was clear: create smaller, more mobile command and control elements that could deploy forward with Marine littoral forces. The reality, as Astacio notes, proved more complex than changing organizational charts. Achieving genuine mobility and survivability required not just organizational redesign but fundamental technological transformation.

The most significant technological breakthrough in Astacio's experience came with the fielding of the Ground/Air Task Oriented Radar (G/ATOR), developed by Northrop Grumman. The impact was immediate and dramatic.

"We replaced five radars with one radar," Astacio states. This single change revolutionized the logistics and operational footprint of Marine air command and control. Instead of multiple separate systems requiring extensive lift assets to transport and complex infrastructure to support, Marines could deploy a single multi-mission radar system capable of air surveillance, air defense, and counter-battery missions.

The mobility advantages extended beyond just consolidating equipment. A G/ATOR system can be transported by C-130 and potentially carried by the new CH-53K with its enhanced lift capacity and stability. Where previous systems required convoy transport and extended setup times, G/ATOR can deploy rapidly to austere locations.

Returning to the discussion, Lieutenant Colonel Astacio underscored that equally important was the fact that the quality and range of data G/ATOR provides surpassed previous systems. The radar's ability to track multiple targets simultaneously while providing precision data enabled new operational concepts. However, Astacio acknowledges that "we still have some issues we need to work with that" regarding range optimization, noting the community has identified pathways forward.

The transition wasn't seamless. Early fielding revealed substantial challenges that offer lessons for any major technology introduction. Marine maintainers trained on the previous AN/TPS-59 radar faced a steep learning curve with the new system. "Trial by fire," Astacio describes it. Maintenance rates were initially poor as Marines learned to maintain and operate the sophisticated new equipment.

These problems highlight broader Department of Defense challenges with parts sustainability for new systems. "Expeditionary logistics is a real thing," Astacio emphasizes. If the operational concept requires positioning radars throughout an area of operations in contested environments, the sustainment enterprise must support that reality.

Eight years into the G/ATOR program, reliability has improved substantially. "We are better than we were three years ago," The trajectory is positive, demonstrating both the transformative potential of the technology and the time required to fully integrate new capabilities.

The F-35 Lightning II represents another technological game-changer for Marine command and control, The aircraft's multi-waveform sensor capabilities and data fusion make it, in effect, a mobile airborne command and control node.

"Just having another mobile, airborne sensor to provide target quality data, that's another game changer," Astacio notes. The F-35 has sensors that can detect and track threats, providing the ground-based command and control system with additional eyes in the sky.

The fundamental issue facing C2 is being able to manage the complete data flow from these systems from start to finish. The challenge is to ensure that the sensors that pick up targets need to be able to pass that to any relevant shooter, a challenge yet to be fully mastered.

This integration gap has significant operational implications. The Marine Corps has invested heavily in both G/ATOR and F-35 precisely because they promise revolutionary command and control capabilities. But if data can't flow seamlessly between systems to enable rapid targeting and engagement, the full potential remains unrealized

The challenge isn't unique to the Marine Corps. Across the joint force and among allies, achieving true multi-domain integration requires overcoming classification barriers, developing common data standards, and creating operational procedures that enable rapid targeting cycles. The technical capabilities exist; the integration framework remains incomplete.

"We have to figure out how to do remote radar, which means I can position a radar forward," Astacio explains. This requires communications over transmission protocols rather than hardwired connections, allowing the command post to operate from a more protected location while sensors push into the area of interest.

The key to making forward sensors survivable is coupling an active radar with passive detection systems. "Now you couple that radar with a passive system," Astacio suggests. The passive system provides continuous surveillance without emitting electromagnetic energy that would identify the sensor location, while the active radar can be selectively employed when needed. "The passive data they're creating rivals a lot of the active in terms of target quality."

This distributed sensing model creates what Astacio calls "a targeting nightmare for the adversary." Instead of a single command post to locate and strike, the adversary faces multiple nodes that may or may not be actual command locations. Some might be sensors, others might be decoys, and determining which merits striking becomes significantly more complex.

Given my work on maritime autonomous systems, it would seem to me that such systems offer potential platforms for these forward sensors. A stable autonomous vessel could carry a passive sensor package, operating for extended periods without exposing crewed platforms to risk. The sensor's passive nature means it's difficult to detect, and even if discovered, destroying an unmanned platform is less operationally significant than eliminating a crewed command post.

Survivability requires more than just distribution and deception. Astacio acknowledges the need for point defense capabilities to protect command and control nodes from incoming fires, whether rockets, missiles, or unmanned systems. The Marine Corps is fielding a prototype solution: the Medium Range Intercept Capability (MRIC), essentially the U.S. version of Israel's Iron Dome system. The system fires interceptor missiles to destroy incoming threats, and it uses G/ATOR as its sensor. "The G/ATOR will really play that function," Astacio confirms.

Even with more mobile equipment, the logistics of deploying command and control capabilities remain substantial. Astacio estimates that moving a Multi-Functional Forward Operating Base or similar command post requires several C-130 sorties to transport power equipment, radars, C2 systems, tentage, and supporting gear.

C-130s, however, are Marine Air-Ground Task Force assets with competing demands across the theater. Astacio frequently requests joint lift assets – C-17s or C-5s – to supplement organic Marine airlift.

This dependency on joint or theater assets limits operational flexibility and requires advance coordination that may not be feasible in crisis response scenarios.

The CH-53K's enhanced lift capacity may offer new options. While its ability to sling-load G/ATOR remains to be fully demonstrated, the aircraft's stability during hover operations could make it viable for transporting sensitive radar components that previous helicopters couldn't safely carry. These incremental improvements in deployment options enhance operational flexibility and reduce dependency on fixed-wing airlift.

Astacio's assessment highlights that the Marine Corps has made substantial progress in creating more mobile, capable command and control capabilities. G/ATOR represents a genuine revolution in Marine air surveillance and control. F-35 offers unprecedented sensor capabilities. Force Design 2030 organizational changes align structure with new operational concepts.

Most fundamentally, the transition Astacio describes, from thinking in terms of bases to thinking in terms of hubs with distributed nodes, requires conceptual innovation as much as technological capability. The platforms and systems largely exist or are in development. What's needed is imagination about how to employ them in ways that create dilemmas for adversaries rather than vulnerabilities for Marines.

"We're heading in the right direction," Astacio concludes. The baseline exists. The path forward is visible.

Lieutenant Colonel Astacio and Marines like him are working to ensure that when the nation needs expeditionary command and control capabilities in contested environments, the Marine Corps will be ready with solutions that are mobile, survivable, and effective. Their progress over the past decade suggests optimism is warranted, even as significant work remains.

CHAPTER 11

COMMAND AND CONTROL IN TRANSITION: OBSERVATIONS FROM THE WING OPERATIONS COMBAT CENTER DURING STEEL KNIGHT

Watching Steel Knight 2025 from inside the Wing Operations Combat Center reveals how C2 is actually changing under load: who really makes decisions, how information flows, and where bottlenecks appear. This chapter offers a cockpit-view of command posts wrestling with distributed operations, kill-web constructs, and coalition visibility in real time. It closes Part Four by drawing out the C2 lessons that feed directly into the impact-force concept developed in the conclusion, linking Steel Knight's experimentation back to the architecture of adaptive power.

During a recent visit to the Wing Operations Combat Center (WOCC) at 3rd Marine Aircraft Wing during the Steel Knight 2025 exercise, the fundamental questions facing Marine Corps aviation command and control came into sharp focus. The conversations with watch officers and operational staff revealed an organization wrestling with perhaps the most critical challenge of modern warfare: how to balance centralized command capabilities with the imperative for distributed operations in an increasingly contested environment?

The WOCC served as the primary command center for the exercise, but the real story was not about the center itself. Rather, it was about the evolving relationship between this centralized node and the distributed command posts being deployed with the actual engagement forces. This dynamic represents the Marines working through a fundamental shift in how command and control must function when facing peer or near-peer competitors.

THE DISTRIBUTED COMMAND CHALLENGE

Since approximately 2018, the Marine Corps and the Department of Defense now rebranded as the Department of War have been fundamentally rethinking what operations in a peer competitor environment actually mean. The traditional model of command and control, built on assumptions of information dominance and relatively permissive operating environments, no longer holds. What has become increasingly clear is that "peer competitor" does not simply mean China or Russia operating at strategic scales. It means that even regional forces and non-state actors now possess capabilities that would have been unthinkable for third-world nations just a generation ago.

Consider the contrast with Marine operations in Lebanon under the Reagan administration. Despite the tragic losses, including the barracks bombing, there was a fundamental asymmetry in technological and operational capabilities. The Marines could generally assume technological superiority in most engagements. That assumption is dead.

Today's Marines cannot enter any operational environment, whether confronting drug cartels or regional military forces, with confidence that they hold the technological edge. The adversary may well possess advanced sensors, precision weapons, cyber capabilities, and integrated air defenses that make traditional command and control structures vulnerable or obsolete.

This reality drives the central question being explored in exercises like Steel Knight: What is the proper relationship between central and distributed command operating centers? What capabilities need to reside with the distributed engagement forces operating forward, and what functions should remain at more centralized, presumably more secure, command posts?

This is not merely an organizational chart exercise. It is about determining what information, what decision-making authority, what communication capabilities, and what situational awareness tools must be pushed to the units that will make contact with the enemy?

THE HUMAN DIMENSION OF COMMAND

Technology plays an obvious role in this transformation, but the discussions at the WOCC highlighted something often overlooked by civilian technologists and even some military planners: the metal furniture matters more than the hardware. The individual Marines working these command-and-control problems are not simply operators of systems; they are the critical variable determining what works and what fails.

One of the watch officers emphasized this point about individual Marine capabilities in the command-and-control environment. Steel Knight is not fundamentally about testing technology, though technology testing certainly occurs. It is about discovering what Marines are actually capable of doing with available tools, identifying where failures occur, and understanding what those failures reveal about the gap between technological promise and operational reality.

This distinction is crucial. The history of military transformation is littered with technologies that looked revolutionary in concept but proved dysfunctional in practice because they could not be effectively integrated into how actual units operate. Technology that is not embedded in Marines' CONOPS (concepts of operations), that Marines do not know how to execute effectively, can reduce capability rather than enhance it. It can create confusion, slow decision cycles, and introduce failure modes that did not exist with previous systems.

This is not a criticism of Marines or any military service. It is simply recognition that technological capability exists only when it is wielded by trained humans who understand both the capabilities and limitations of their tools. Civilian technologists often project what a technology could mean without grasping what it actually takes for real people, under stress, in degraded conditions, to make that technology deliver value rather than confusion.

LEARNING FROM THE
OSPREY EXPERIENCE

The V-22 Osprey provides an instructive parallel to the current command and control evolution. The Osprey's transformation from a controversial acquisition program to an essential Marine Corps capability did not happen overnight or simply because the aircraft's technical problems were solved. It required years of Marines learning how to employ the aircraft effectively, discovering what it could and could not do, and developing new operational concepts that exploited its unique capabilities.

I have published two books on the Osprey transformation which significantly documents this process of what real world transformation looks like.

Every three or four years, Marines developed new ways to use the Osprey that expanded what the platform made possible. This was not because each generation of Marines was smarter than the last. It was because they had more experience with the aircraft, they had learned from previous operations and exercises, and they had time to discover both the good and the bad in the platform. The Marines in 2025 understand things about Osprey employment that the Marines in 2007 could not have known because they did not yet have eighteen years of operational experience.

Every technology brings good, bad, and ugly. There are things it does superbly, things it cannot do despite hopes or expectations, and things it absolutely should not be used for. The first Marines to push the Osprey beyond the artificial 200-mile operational box for an ARG-MEU to exploit its true 1,000-mile range were not following doctrine; they were learning what was actually possible and developing new concepts based on experience rather than theory.

The same learning process is now occurring with command-and-control systems and concepts. What can distributed nodes actually accomplish? What information can they generate and transmit? What decisions can they make with available data? What happens when communications degrade or fail?

These questions can only be answered through repeated exercises, honest assessment of failures, and gradual refinement of both technology and tactics.

THE UKRAINE WAR'S LESSONS

The ongoing war in Ukraine has provided stark lessons about the pace of tactical and operational adaptation. One of the critical takeaways is the concept of hybrid operations, in which forces constantly recombine capabilities in new patterns to defeat enemy countermeasures and exploit emerging opportunities. The Ukrainians might employ HIMARS and drones in a particular pattern one month, then shift to an entirely different combination and employment concept the next month to maintain effectiveness and surprise.

This pattern of rapid adaptation cannot be assumed to be a Western or Ukrainian advantage. Adversaries globally are demonstrating similar capabilities for fast cycle learning and adaptation. This has profound implications for command and control. The engagement force on the ground is encountering enemy systems, tactics, and capabilities in real-time. They may be facing weapon combinations or employment patterns that were not present in intelligence preparations. They may discover that an enemy capability assessed as limited is actually being employed with unexpected sophistication.

This information from the engagement force about what they are actually encountering is crucial and cannot be fully captured by space-based sensors or other remote intelligence collection. Human contact, human observation, and human judgment about what is actually happening in the tactical environment provide irreplaceable value. The engagement force must be able to communicate what they are experiencing back to higher commands rapidly enough for that information to influence operations.

But this creates tension with distribution and survivability. The more communications occur, the more electronic signatures are created that sophisticated adversaries can exploit. The more centralized the command node, the more lucrative a target it becomes. The more distributed the command structure, the harder it becomes to maintain common operational pictures and synchronized operations. These tensions cannot be resolved through technology alone. They require careful thinking about what information actually needs to flow where, what decisions can be made at what levels, and what risks are acceptable in exchange for what capabilities.

WORKING THE HYBRID BALANCE

The discussions at the WOCC kept returning to questions about the correct hybrid balance. What should go forward with the distributed command posts? What must remain at the rear command post?

This is not a simple division of strategic versus tactical decision-making. It involves questions about which technical capabilities can be sustained forward, what bandwidth is realistically available, how much redundancy is needed, and what happens when the distributed nodes lose connectivity with rear commands.

The attack squadron Captain working in the WOCC spoke to exactly these challenges. The current approach represents a hybrid model with some capabilities forward and others remaining at the main post. But the challenge is determining what this hybrid structure actually enables and where it creates new problems or vulnerabilities. Does having certain capabilities forward provide genuine operational advantage, or does it simply create additional nodes that require protection and sustainment? What does the rear command post provide that the forward posts cannot duplicate, and is that division of capability optimal or merely traditional?

These questions do not have simple answers, and different operational scenarios may demand different solutions. A Marine Expeditionary Unit operating from amphibious ships faces different constraints than a Marine littoral regiment operating from expeditionary advanced bases. A crisis response mission in a permissive environment differs from contested operations against a sophisticated adversary. The hybrid balance must be contextual rather than doctrinal.

TECHNOLOGY INTEGRATION VERSUS OPERATIONAL REALITY

One of the persistent themes in the WOCC conversations was the gap between what technology promises and what it actually delivers in operational contexts. Technologists and acquisition professionals often focus on what systems can do under ideal conditions: bandwidth, processing speed, range, resolution. But operational effectiveness depends on what systems can do when employed by Marines who are tired, under stress, operating in degraded conditions, with limited training time, and facing unexpected challenges.

This is why exercises like Steel Knight are essential. They create opportunities to discover where technology fails to deliver, where Marine training is insufficient, where concepts prove unworkable, and where unanticipated interactions between systems create problems. These discoveries are not failures of the exercise; they are the exercise's purpose. Better to find these problems in training than in combat.

The WOCC staff emphasized that they are testing not just whether technology works, but whether Marines can use it effectively to accomplish actual missions. Can they maintain situational awareness with available systems? Can they coordinate fires and maneuver across distributed nodes? Can they adapt when communications fail or when the enemy does something unexpected? These are questions about human-machine integration, not just technical specifications.

THE PATH FORWARD

The transformation of Marine Corps command and control for distributed operations in contested environments represents a journey rather than a destination. There is no final answer to the question of how centralized versus distributed command structures should be configured. Instead, there is a continuous process of adaptation as technology evolves, as adversaries develop new capabilities, as Marines gain experience, and as new operational concepts emerge.

What the visit to the WOCC during Steel Knight revealed is an organization taking this challenge seriously, asking hard questions, testing assumptions, and trying to learn from both successes and failures. The conversations with watch officers and operational staff showed Marines thinking deeply about not just how to employ new technology, but about the fundamental nature of command and control in modern warfare.

The relationship between centralized command posts like the WOCC and distributed forward nodes is still being worked out. The correct hybrid balance remains elusive and may vary by mission and environment. The integration of technology with Marine capabilities continues to evolve. But the willingness to exercise these problems, to discover where current approaches fail, and to adapt based on experience represents exactly the kind of institutional learning that will determine whether the Marine Corps can successfully operate in the contested environments it will inevitably face.

The challenge is not to find perfect solutions but to develop adaptive organizations that can learn faster than their adversaries, that can integrate new capabilities effectively, and that can distribute decision-making without losing coherence. The work happening at the WOCC and in exercises like Steel Knight represents essential steps in that ongoing transformation. The Marines are working the problems, discovering what works and what does not, and gradually building the capabilities and concepts that will define command and control in future operations.

PART THREE
WORKING THE TRANSFORMATION

Part III, "Working the Transformation," shifts the book from description to evaluation, asking how far the Marine Corps has actually progressed toward becoming an impact force rather than simply a redesigned crisis-response organization.

This section treats transformation as a lived process, not a finished product, and uses aviation plans, logistics realities, and Force Design updates to gauge whether the Corps can generate disproportionate effects for the joint and allied team under conditions of contested logistics, degraded communications, and strategic ambiguity.

At the center of Part III is Marine Corps aviation, now cast as connective tissue for distributed operations rather than a collection of mission-segmented platforms. The 2025–2026 Marine Aviation Plans and Project Eagle's three-horizon construct (Fight Tonight 2026–2030, Bridge the Gap 2031–2035, Future Fight 2036–2040) show an enterprise deliberately moving from platform-centric to data-centric, decision-centric aviation.

Platforms like the F-35, MQ-9, H-1, KC-130J, V-22, and CH-53K are assessed less for their individual firepower and more for their ability to act as airborne brains, sensors, tankers, and connectors that sustain a distributed kill web. Aviation Ground Support is elevated to the seventh function of Marine aviation precisely because fuel, power, and maintenance at austere, shifting locations are recognized as the non-negotiable backbone of Distributed Aviation Operations.

Steel Knight 25 provides the primary laboratory for testing whether these concepts work in practice. Part III examines how I MEF treated the exercise as a campaign, not a scripted readiness drill, assuming no sanctuary, no permissive buildup, and constant surveillance from the opening hours. Concepts such as hub-spoke-node architectures, time-stamped nodes with inherent "shelf lives," and the Heavy Node problem are explored in detail to show how signature, survivability, and tempo interact in a real kill-web fight. The exercise's role in certifying Marine Rotational Force–Darwin links experimentation directly to Indo-Pacific realities, reinforcing that impact force design must work against a peer adversary, not just on PowerPoint.

A recurring theme in Part III is that logistics is the limiting factor for distributed operations. The sections on sustainment detail how range, refueling, and maintenance quickly become pacing functions when V-22s, KC-130Js, and CH-53Ks operate at the outer edges of their envelopes, often in electromagnetic silence. Commanders' reflections highlight the difficulty of building the right forward maintenance packages, the immaturity of additive manufacturing for aviation, and the gap between predictive maintenance promise and naval supply-chain reality. The CH-53K, with its lift and diagnostics, is presented as a potential game-changer, but only if acquired and supported in sufficient depth to reduce over-reliance on KC-130s and joint lift "magic."

Part III also delves into the cultural and command-and-control dimension, arguing that the primary risk to an impact force is not technology but centralization under stress. Rich digital interoperability means that generals and forward captains now see much of the same data, creating a "delegation dilemma" where higher headquarters may be tempted to reach down and direct fleeting engagements, neutralizing the kill web's speed advantage.

Mission command and trust at the edge are indispensable: if distributed nodes must wait for permission to act on a 32-second transmission, adversaries will kill the node before the joint system can respond. In this sense, Ukraine is treated as a proxy conflict under-scoring that adaptation speed and decision advantage matter more than raw platform performance or notional order of battle.

Finally, Part III uses the October 2025 Force Design Update as an indicator of whether the Corps is genuinely learning. Decisions such as retaining 4th Marines as a conventional infantry regiment and restoring breaching and bridging capabilities divested with tanks are presented as evidence-based reversals driven by exercise-derived lessons, not ideological course corrections.

Across its chapters, Part III portrays a Marine Corps that has moved beyond the "inside force equals ship-killer" mindset toward a more demanding role as joint JTAC for the kill web, using small, survivable units to find, fix, and enable fires from deeper magazines across the joint and allied force.

MARINE CORPS AVIATION TRANSFORMATION: A COMPREHENSIVE ANALYSIS OF THE 2025-2026 AVIATION PLANS

The United States Marine Corps stands at a critical inflection point in its aviation modernization journey. The 2026 Marine Aviation Plan represents not merely an incremental update to the previous year's strategy, but rather a fundamental acceleration of the service's transformation toward a distributed, data-enabled, and decisively lethal Aviation Combat Element (ACE).

This chapter examines the evolution between the 2025 and 2026 plans, revealing how the Marine Corps is balancing immediate readiness requirements with aggressive modernization imperatives designed to outpace peer competitors in contested environments.

STRATEGIC FRAMEWORK EVOLUTION: FROM FORCE DESIGN TO PROJECT EAGLE

The 2025 Aviation Plan operated within the broader Force Design 2030 framework, emphasizing expeditionary operations and distributed lethality. While these concepts remain foundational, the 2026 plan introduces Project Eagle as the strategic blueprint specifically for

Marine Aviation. This represents a significant maturation in how the service articulates its aviation transformation.

Project Eagle extends planning horizons across three Future Years Defense Programs (FYDPs), creating a structured approach that translates operational requirements into deliberate, long-term modernization. This framework addresses a persistent challenge in defense acquisition: the tendency to optimize for near-term budget cycles at the expense of strategic coherence. By extending the planning aperture to encompass FYDP-1 (2026-2030: "Fight Tonight"), FYDP-2 (2031-2035: "Bridge the Gap"), and FYDP-3 (2036-2040: "Future Fight"), Project Eagle ensures that aviation capabilities evolve at the speed of relevance rather than the pace of bureaucratic processes.

Central to the 2026 plan is the T-E-A-M framework, which serves as both an operational philosophy and an organizational culture statement:

- Take care of our Marines, Sailors, and aircrew.
- Execute the Basics with Brilliance and precision.
- Attain and Maintain our Mission Readiness.
- Mitigate Risk in everything we do.

This framework represents more than aspirational language. It acknowledges that technological superiority alone cannot guarantee success: the human dimension remains decisive. The emphasis on executing basics with brilliance addresses a sobering reality revealed in mishap analysis: 78.8% of aviation mishaps involve human factors, with nearly 30% attributable to procedural non-compliance. The ambitious "26 in 26" safety initiative, aiming to reduce total Class A-D mishaps by 26 in fiscal year 2026, demonstrates how the T-E-A-M philosophy translates into measurable outcomes.

OPERATIONAL CONCEPTS: DISTRIBUTED AVIATION OPERATIONS AND DECISION-CENTRIC WARFARE

Both plans emphasize distributed operations, but the 2026 plan provides substantially greater clarity on implementation. DAO is explicitly defined as "Marine Aviation's central contribution to Expeditionary Advanced Base Operations (EABO) and the Corps' ability to operate inside contested maritime spaces." This operational concept disperses aviation assets across multiple expeditionary sites to increase survivability, complicate adversary targeting, and provide persistent support to the Stand-In Force.

The 2026 plan goes beyond conceptual descriptions to address the practical enablers of distributed operations. It identifies Aviation Ground Support (AGS) as the "backbone of DAO," elevating it to the seventh function of Marine Aviation. This organizational change recognizes that as the ACE becomes more distributed and expeditionary, AGS becomes the critical determinant of whether Marine Aviation can sustain combat power in contested environments. The plan directs comprehensive recapitalization of AGS capabilities, ensuring the ACE can generate fuel, power, maintenance, mobility, airfield services, and expeditionary infrastructure from austere and rapidly shifting locations.

Perhaps the most significant conceptual advancement in the 2026 plan is the formal introduction of Decision-Centric Aviation Operations. DCAO enhances DAO by increasing the speed and quality of decision-making through data, digital tools, and emerging AI/ML-enabled decision support. Positioned as the aviation-specific application of the service's Project Dynamis initiative, DCAO connects Marine Aviation transformation to the broader Department of Defense effort to build an AI-powered, decision-centric warfighting network.

DCAO addresses a fundamental challenge of distributed operations: maintaining decision advantage when forces are dispersed across vast distances and operating under communications constraints. By achieving "data dominance, the ability to sense, process, share, and act on information faster than an adversary", DCAO ensures distributed sites remain connected, informed, and decision-relevant. As the capability matures, it will enable distributed forces to rapidly re-task, adapt to changing conditions, and close kill webs faster than any competitor.

PLATFORM MODERNIZATION: ACCELERATING TRANSITIONS AND EXPANDING CAPABILITIES

Both plans detail the ongoing transition to the F-35, but the 2026 plan reflects a force moving beyond initial operational capability toward full integration. By the end of 2026, the Marine Corps will have received 205 F-35B and 56 F-35C aircraft, supporting 14 fleet squadrons plus developmental/operational test units and Fleet Replacement Squadrons on both coasts.

A significant development detailed in the 2026 plan is the decision to increase all fleet F-35 squadrons from 10 to 12 Primary Aircraft Available (PAA). This expansion, with aircraft increases beginning in FY30 and manpower changes implemented in FY28, reflects operational analysis demonstrating that larger squadrons enhance deployment flexibility and maintenance capacity. The plan projects the total inventory will reach 313 aircraft to support this expanded structure.

The 2026 plan also provides greater specificity on F-35 modernization priorities. Technical Refresh-3 (TR-3) aircraft continue delivery, providing the hardware foundation for future Block 4 capabilities. Critical weapon integrations receive detailed attention, with the fielding of software builds 30P08 and 40P02 bringing Small Diameter Bomb II to the fleet. The plan emphasizes accelerating integration of AGM-88G AARGM-ER, the AGM-158 family (JASSM-ER and LRASM), AIM-9X Block II+, and Six-in-the-bay configurations for the F-35C.

Perhaps most significantly, the 2026 plan positions the Marine Corps as leading the integration of Collaborative Combat Aircraft with the F-35. The MAGTF Unmanned Expeditionary TACAIR (MUX-TACAIR) program will increase F-35 survivability and lethality, enabling successful mission execution across a wide range of developing threat environments.

The F/A-18 Final Chapter

While both plans acknowledge the F/A-18 Hornet's approaching sunset, the 2026 plan provides precise timelines and demonstrates the service's commitment to sustaining capability through transition. The Hornet will continue supporting Unit Deployment Program requirements through the end of FY28, with VMFA-323's dual-mission role as both operational squadron and Fleet Replacement Squadron consolidated into an FRS function in FY26.

The 2026 plan details aggressive final-fit modernization initiatives designed to maximize Hornet relevance and survivability through sundown. The completion of AN/APG-79(v)4 Active Electronically Scanned Array radar fielding by FY26 represents a quantum leap from mechanically scanned radar technology. Combined with a modern electronic warfare suite, the pursuit of net-enabled extended-range weapons, beyond-line-of-sight communications, and Auto Ground Collision Avoidance System integration, these enhancements significantly improve the Hornet's readiness, lethality, and survivability for its remaining service life.

AV-8B Harrier: Honoring a Legacy

The 2026 plan provides comprehensive details on the AV-8B Harrier's sunset, including formal commemoration events scheduled for June 2026 at MCAS Cherry Point. VMA-223's final operational detachment with the 22nd MEU represents the culmination of decades of service. The plan explicitly addresses the transition pathway for Harrier personnel, with highly experienced aircrew and maintenance personnel primarily transitioning to the F-35B, ensuring their expertise continues benefiting Marine Aviation.

Tiltrotor Aviation: MV-22 Sustainment and Modernization

Both plans address MV-22 operations, but the 2026 plan provides significantly greater detail on initiatives to enhance reliability, safety, and operational capability. The Proprotor Gearbox improvements represent the most visible safety-focused initiative, with the introduction of the -119 PRGB (using Triple-Melt steel for critical components) in June 2025, followed by the enhanced -123 PRGB in January 2026. The plan projects an unrestricted operational fleet by December 2027 and complete transition to -123 gearboxes by January 2033.

Beyond safety improvements, the 2026 plan details ambitious modernization initiatives. The Osprey Drive System Safety and Health Instrumentation (ODSSHI) program will incorporate sensors in critical areas of the PRGB and drivetrain, providing vibration signature data that enables proactive maintenance by forecasting component failures. The V-22 Fleet Optimization and Reduction in Configuration Effort (VFORCE) initiative will convert Block B airframes to Block C-Mission Computer Obsolescence Initiative configuration, reducing unique configurations and standardizing aircraft capabilities across the fleet.

The 2026 plan also addresses a critical operational limitation: the pivot from Tailored Nacelle Improvements to full Nacelle Improvements, replacing entire nacelle assemblies rather than focusing solely on wiring bundles. This decision, informed by validated performance data from Air Force Special Operations Command V-22 operations, promises substantial readiness gains that single-system improvements could not achieve.

Heavy Lift: CH-53E to CH-53K Transition

The CH-53K transition continues as a centerpiece of both plans, but the 2026 plan provides enhanced clarity on deployment timelines and operational capability milestones. HMH-461, the first operational squadron, achieved its full complement of 16 aircraft in FY26, transitioning from 0.75 to 1.0 unit status. The plan projects the first CH-53K MEU detachment will CHOP (Change of Operational Control) in FY26, preparing for its first operational deployment in FY27.

The 2026 plan explicitly addresses the complex sequencing required for MARFORPAC Echo-to-Kilo transition, which requires sufficient aircraft inventory to support MARFORCOM requirements and a delivery schedule capable of sustaining consecutive transitions. The planned deployment order, West Coast MEUs (11th, 13th, 15th), then 31st MEU, then UDP, ensures maximum HMH capacity is maintained with minimal impact on deploy-to-dwell ratios.

Significantly, the 2026 plan identifies three "pillars" supporting CH-53K deployment decisions: Aircraft Inventory, Sustainment (Spares and Repairs), and Aircraft Capability. This framework provides transparency into the decision-making process and helps fleet operators understand the interdependencies that govern transition timelines.

Attack and Utility Helicopters: H-1 Mid-Life Modernization

Both plans address H-1 modernization, but the 2026 plan provides substantially greater detail on the Structural and Power Improvement for Next-gen Effects (SPINE) program and its integration with advanced weapons and survivability systems. SPINE provides greater electrical power capacity to expand current warfighting capabilities and enable integration of future weapons including Precision Attack Strike Munition, AIM-9X, and counter-unmanned aerial systems capabilities.

The 2026 plan also details initiatives to grow the two East Coast squadrons from 0.75 to full 1.0 squadrons by the end of FY31, increasing total aircraft inventory to 313 H-1 aircraft and achieving 6.0 squadron equivalents. This expansion addresses a persistent shortfall in East Coast assault support capacity.

DIGITAL TRANSFORMATION: THE FOUNDATION OF DECISION ADVANTAGE

While the 2025 plan acknowledged digital interoperability importance, the 2026 plan positions DI/MAGTF Agile Network Gateway Link (MANGL) as fundamental to achieving decision advantage in distributed operations. The plan articulates a clear vision: providing information to users at the right time to successfully engage adversaries and improve operational efficiency during conflict and competition.

The 2026 plan details the "four pillars of DI", sensors, processors, interfaces, and radios/apertures, that platforms must integrate to reliably exchange relevant information. It outlines modernization strategy to update architecture and align with Combined Joint All-Domain Command and Control (CJADC2) architecture while maintaining alignment with waveform advancements and other service investments.

Significantly, the 2026 plan details fleet employment initiatives that extend DI/MANGL capabilities beyond aircraft. The Extended Tactical Network enables connections and mission synchronization across multiple theaters, supporting ChatSurfer communications and Tactical Assault Kit server connections. MEU Landing Force Operations Centers "Fly away kits" and tactical vehicles equipped with DI kits enable situational awareness of command operations centers and dismounted forces.

The MV-22 serves as the lead platform for DI/MANGL, with a capability demonstration supported by PMA-275 scheduled for early FY26 and platform fielding decision expected in FY27. The plan details integration efforts for KC-130J and CH-53K, while noting the H-1 program is integrating and fielding DI systems through PMA-276.

ARTIFICIAL INTELLIGENCE AND MACHINE LEARNING IN AVIATION SUSTAINMENT

The 2026 plan introduces a comprehensive framework for "Transforming Aviation Sustainment for the Future Fight" through AI/ML integration. This initiative directly addresses the problem statement: "Marine Corps Aviation remains reactive in maintenance, supply, and operations planning, limiting readiness and reducing the ability to sustain distributed aviation operations and crisis response."

The plan organizes AI/ML integration into three Lines of Operation:

- Line of Operation 1: Dynamic Aviation Supply focuses on revising supply package structures to provide sufficient depth for highly dynamic, nodal webs of aircraft and support sites. The initiative evolves Supplemental Aviation Spares Support employment and leverages AI/ML capabilities to review support packages, capturing latest configurations and failure rates to improve deployed aircraft readiness rates while reducing costs and material footprint.
- Line of Operation 2: Predictive Maintenance represents a fundamental transition from reactive to proactive maintenance culture. The plan details partnerships with industry to explore Advanced Maintenance Training Academies that leverage manufacturer instructors and engineering specialists, providing hands-on experience and 3D courseware for critical maintenance tasks on H-1 and V-22 aircraft. These programs establish a strong baseline for Maintenance Chiefs while fostering a culture of maintenance excellence.
- Line of Operation 3: Optimized Operations expands data-centric warfighting across the full spectrum of aviation activities by fusing data from historically separate systems like NALCOMIS, M-SHARP, and GCSS-MC. The plan describes developing a suite of AI-enabled tools to automate and optimize complex, data-intensive tasks of scheduling and managing flight operations and maintenance. The ultimate outcome is a direct increase in combat readiness and lethality by generating safer, more efficient operations that reduce unscheduled downtime and increase aircraft availability.

WEAPONS AND ELECTRONIC WARFARE: EXTENDING REACH AND ENHANCING LETHALITY

Both plans emphasize long-range precision strike, but the 2026 plan provides specific timelines and integration details. The AGM-158C

LRASM C-1 variant achieved operational fielding on F/A-18E/F in FY24, with integration continuing on F-35B and F-35C. The production line will cut in the upgraded C-3 variant beginning in FY26, providing even more capability within the FYDP.

The 2026 plan also details continued development and integration of stand-off and net-enabled weapons on the F-35, with weapons acceleration efforts fielding highest-priority capabilities to operational commands quickly while providing maximum envelope and operational flexibility. Net Enabled Weapons University will help meet increased demand for subject matter experts to assist in holistic squadron training for maintainers and operators.

Rotary-Wing Precision Munitions

A significant development detailed in the 2026 plan is the formal initiation of the Precision Attack Strike Munition (PASM) Program of Record, transitioning from the Long-Range Attack Missile Defense Innovation Acceleration. The LRAM DIA completed its Operational Demonstration in September 2025 with successful employment from an HX-21 AH-1Z. Efforts now focus on fielding capability as quickly as possible to deployed squadrons during FY27.

The plan also details continued development of the Advanced Precision Kill Weapon System, with efforts to improve APKWS in conjunction with Navy and Air Force. These enhancements, already demonstrated by Marine H-1s against various air and ground targets, significantly improve ability to provide offensive and defensive air support.

Electronic Warfare Modernization

While both plans address electronic warfare, the 2026 plan provides substantially greater detail on the Intrepid Tiger II EW Family of Systems and future development pathways. Current capabilities include IT-II Versions (V)1 and (V)3 deployed on AV-8B and UH-1Y respectively, with (V)4 in developmental test on MV-22B targeting IOC in FY27. IT-II Block 5 incorporates Counter-Radar capabilities with priority platforms being UH-1Y, AH-1Z, MV-22, and KC-130.

The 2026 plan details ambitious future initiatives including Common Architecture for Persistent EW (CAPE), which allows operationalized capabilities such as Multi-Domain Strike, and development of MUX TACAIR EW capabilities to increase survivability and lethality of manned platforms. A Common Framework Environment will deliver flexible government-owned software architecture for EW, geolocation, waveform, SIGINT, and other electromagnetic spectrum operations capabilities supporting multiple platforms and vendors.

Ground-Based Air Defense: Capability Acceleration

Both plans address Marine Air Defense Integrated System fielding, but the 2026 plan reflects accelerated delivery and expanded force structure. The plan projects 190 total MADIS systems with IOC in FY25 and FOC in FY33, providing foundational capability for Low Altitude Air Defense Battalions and Littoral Anti-Air Battalions.

The 2026 plan details the MADIS C-UAS Engagement System, which enhances lethality by integrating an autonomous effector for Group 1-2 UAS and an economical effector for Group 2-3 UAS and fixed- or rotary-wing threats. Future efforts focus on integrating CES and MADIS subcomponents onto unmanned ground vehicles, along with investing in emerging technologies such as directed energy, low-cost kinetic effectors, and the Army's Next Generation Short Range Interceptor.

Medium Range Intercept Capability

The 2026 plan provides comprehensive details on Medium Range Intercept Capability fielding, with 3 Batteries comprising 48 launchers achieving IOC in FY26 and FOC in FY30. MRIC complements MADIS and L-MADIS by extending engagement options to advanced threats, integrating with Marine Corps radars and C2 systems to provide mobile and expeditionary medium-range defense.

ORGANIZATIONAL CHANGES

A significant development detailed in the 2026 plan is the reactivation of 4th LAAD in FY29 with an intended FOC of 12 MADIS systems in FY33. This addresses the longstanding gap of LAAD capability in reserves to augment active component and catch talented LAAD Marines leaving the force.

Aviation Expeditionary Enablers: Restructuring for Distributed Operations

The 2026 plan provides comprehensive details on the transition to Multifunction Air Operations Center, which enhances MAGTF lethality, operational depth, and flexibility compared to legacy air C2 models. The transition plan includes:

- FY26: All Direct Air Support Center and Tactical Air Operations Center Marines consolidate into unified cadre of MAOC Marines.
- FY27-28: Active component MACGs restructure to establish Marine Air Support Squadrons purpose-built for MAOC employment and Marine Air Control Squadrons optimized for air traffic control and Tactical Air Command Center missions.
- FY29: Reserve Component reorganization.

This organizational transformation represents fundamental reconceptualization of how Marine Aviation conducts command and control in distributed environments.

Air Traffic Control Modernization

The 2026 plan details significant air traffic control modernization initiatives. The Expeditionary Airport Surveillance Radar is expected to achieve IOC in FY28, replacing the surveillance function of ATNAVICS while bringing increased capability. Aligned with EASR fielding, efforts are underway for Expeditionary Precision Approach Landing Capability to support operations in contested environments, critical to distributed aviation operations.

MATC has also begun integration of MANGL/MAGTAB within its formations to enable scalable, interoperable situational awareness at remote air sites and Forward Arming and Refueling Points, ensuring air traffic control functions as an effective node within the broader Marine Air Command and Control System.

AVIATION GROUND SUPPORT ELEVATION

As previously noted, the 2026 plan elevates AGS as the seventh function of Marine Aviation, acknowledging its criticality to distributed operations. The plan details comprehensive AGS systems modernization including:

- VTOL Surfacing Systems (52 systems, FOC FY26).
- Aircraft Arresting System Replacement (25 systems, IOC FY29, FOC FY33).
- C-130 Transportable ARFF Apparatus (68 systems, currently unfunded, IOC FY28).
- Tactical Aviation Ground Refueling System (78 systems, FOC FY26).

The plan also identifies future AGS capabilities under development: Rapid Intervention System (compact, MV-22 transportable firefighting), Expeditionary Fuel Truck (C-130 transportable refueling asset), Expeditionary Dust Abatement System (water and soil palliative distribution), and Automated Airfield Damage Assessment System (sUAS equipped with multi-spectral imaging and LiDAR powered by AI-driven analysis).

UNMANNED SYSTEMS: EXPANDING THE FOOTPRINT

While both plans address MQ-9A operations, the 2026 plan reflects a force moving beyond initial operating capability toward comprehensive integration. The detailed capabilities schedule projects:

- VMU-1: 6 AV-25, 6 GCS, 3 Sky Tower 1, 6 Sky Tower 2, 6 pLEO, 6 DAAS, 5 ES.
- VMUT-2: 4 AV-25, 3 GCS, 4 Sky Tower 2, 4 pLEO, 4 DAAS.
- VMU-3: 6 AV-25, 4 GCS, 3 Sky Tower 1, 7 Sky Tower 2, 6 pLEO, 6 DAAS, 5 ES.
- VMX-1: 2 AV-25, 1 GCS, 3 Sky Tower 1, 2 Sky Tower 2, 2 pLEO, 2 DAAS, 1 ES.

This expansion reflects recognition that VMU squadrons operating from expeditionary sites in the Pacific provide the MAGTF with continuous reach and decision advantage. The 2026 plan positions MQ-9A as delivering Maritime Domain Awareness, extending MAGTF C2, and integrating seamlessly into Naval and Joint campaigns.

MUX TACAIR: LEADING EDGE INNOVATION

The 2026 plan details the Marine Corps' leading role in developing and fielding Collaborative Combat Aircraft integrated with F-35. MUX TACAIR Increment I focuses on providing low-cost, risk-worthy capability that enhances F-35 effectiveness in peer/near-peer fights. The Marine Corps successfully demonstrated Manned-Unmanned Teaming between F-35 and XQ-58 during experimentation flights, with upcoming milestones including taxi testing and first flight of the Conventional Takeoff and Landing variant.

The establishment of the MUX TACAIR Transition Task Force in January 2026 demonstrates organizational commitment to fielding this brand-new capability. This represents a significant departure from traditional acquisition approaches, with the Marine Corps willing to accept operational prototypes and iterative development rather than waiting for fully mature systems.

TRAINING AND READINESS: MAINTAINING THE EDGE

While both plans acknowledge MAWTS-1's role, the 2026 plan provides substantially greater detail on how the WTI course has evolved to address future operating environment challenges. The modern course is defined by four key themes: distributed across vast distances (318,000 square miles of airspace), contested by thinking adversaries, conducted across multi-domains, and constantly evolving to meet new threats.

The 2026 plan details how WTI actively incorporates service concepts, ensuring emerging doctrine is tested and refined in realistic settings. It fosters extensive Joint and Coalition partnerships, with regular participation from Navy Third Fleet combatants and Joint Naval Aviation platforms. The capstone event, Strike 4, demonstrates the pinnacle of distributed operations, involving movement of live ordnance via tactical airlift to a FARP where F-35s are hot-refueled and hot-loaded before conducting integrated live-fire night strike against live 4th and 5th Generation adversaries.

Significantly, the 2026 plan positions the WTI course as a critical laboratory for advancing emerging concepts and experimenting with new technologies, in close collaboration with HQMC, MCWL, and industry partners. This commitment to innovation manifests through numerous initiatives: rehearsing complex distributed operations, pioneering tactics for over-the-horizon targeting, advancing sUAS integration for offensive action and integrated counter-UAS development, developing new training to enhance EABO capabilities, honing over-water personnel recovery with MV-22, implementing Aircrew Enroute Casualty Care for distributed forces, and experimenting with MV-22 Anti-Submarine Warfare.

LIVE, VIRTUAL, AND CONSTRUCTIVE TRAINING

The 2026 plan provides comprehensive details on Aviation Distributed Virtual Training Environment expansion. While enhanced capabilities were fielded at 2d and 3d MAWs, FY26 sees expansion to 1st and 4th MAWs to integrate Marine Aviation platforms and systems into a common training environment.

ADVTE connects to Marine Corps Training Enterprise Network and Navy Common Training Environment to support naval training, while demonstrating interoperability with the Air Force's Distributed Mission Operation Network through participation in VIRTUAL FLAG training events. This integration enables collaborative unit-level training and large-force exercises within a unified, scalable, all-domain training environment supporting future operating concepts and Force Design objectives.

RESERVE COMPONENT INTEGRATION: TOTAL FORCE AVIATION

While both plans address reserve component roles, the 2026 plan reflects significant force structure growth and capability expansion within 4th MAW. The reactivation of VMFT-402, VMU-4, 4th LAAD, and VMFA-134, along with relocation and transition of VMFA-112 and growth of HMH-772 to 16 CH-53K aircraft, demonstrates commitment to maintaining robust reserve aviation capabilities.

The 2026 plan provides specific timelines for these activations and details the rationale behind each. VMU-4 reactivation in FY28 provides mission crew augmentation to active component, postures to capture transitioning critical UAS talent, and receives future UAS equipment solutions as service priorities dictate. The 4th LAAD reactivation in FY29 with intended FOC of 12 MADIS systems in FY33 addresses the longstanding gap in reserve component ground-based air defense capability.

ADVERSARY AIRCRAFT MODERNIZATION

The 2026 plan provides comprehensive details on F-5 modernization through the Avionics Reconfiguration and Tactical Modernization for Inventory Standardization (ARTEMIS) program. Repatriated Swiss F-5s undergo structural and avionics upgrades that improve aircraft safety, sustainability, and flyability. Further advancements in Adversary Mission Systems increase situational awareness and capability through encrypted Tactical Combat Training System I and II.

Significantly, the 2026 plan acknowledges the need to pursue "Adversary Next" with the goal of replacing F-5 over the next 10-15 years as the platform reaches end of service life. This forward-looking approach ensures Marine Aviation maintains professional adversary training capability essential for preparing aircrew for 5th Generation threats.

FUTURE ATTACK STRIKE AND THE LONG-RANGE VISION

The 2026 Aviation Plan explicitly addresses the "next generation" transformation of the Aviation Combat Element through Project Eagle's extended planning horizon. Within FYDP-3 (2036-2040), designated as the "Future Fight" timeframe, the Marine Corps is developing Future Attack Strike (FASt) as the eventual replacement for the H-1 platform. According to the plan, FASt capability is being developed to provide long-range fires and Close Air Support to the ground force while serving as a Joint Force kill web enabler.

The program continues to evolve through Weapons Integration Risk Reduction trade studies designed to drive innovation and experimentation, with conceptual solutions being analyzed to inform requirements and acquisition pathways. Enhanced capabilities under consideration include kinetic and non-kinetic launched effects, long-range precision fires, advanced survivability, digital interoperability, and electronic warfare—positioning FASt for an Initial Operational Capability around 2040.

THE OSPREY'S CROSSROADS: VFORCE AND BEYOND

The V-22 modernization strategy represents a critical near-term decision point that will fundamentally shape Marine Aviation's assault support capabilities through mid-century. The plan describes the V-22 Fleet Optimization and Reduction in Configuration Effort (VFORCE) initiative, which will convert Block B airframes to the Block C-Mission Computer Obsolescence Initiative configuration, reducing unique configurations and standardizing capabilities across the fleet. More significantly, the Marine Corps is pursuing additional modernization efforts informed by the Center for Naval Analysis Renewed V-22 Aircraft Modernization Plan (ReVAMP) study to ensure platform relevance and reliability until the end of the V-22 fleet's service life, while simultaneously developing and informing requirements for Next Generation Assault Support aircraft.

The document makes clear that the decisions being made today on Osprey sustainment, including the full-nacelle-replacement Nacelle Improvements initiative, Flight Control Computer redesign, and cockpit technology upgrades through VeCToR, will directly impact the timeline and requirements for NGAS.

BRIDGING CURRENT READINESS WITH FUTURE CAPABILITY

This dual-track approach, sustaining and modernizing current platforms while developing their replacements, exemplifies the fundamental tension the Aviation Plan seeks to resolve: maintaining "crisis response readiness" today while delivering "long-term modernization of Marine Aviation for the future fight."

The plan's investment priorities reflect this balance, with H-1 modernization focusing on lethality (SPINE/PASM), digital interoperability (LINK-16/pLEO), and survivability (DAIRCM) through the 2040s, even as FASt development accelerates. Similarly, MV-22 priorities center on ODSSHI for drivetrain reliability, VFORCE for configuration standardization, and nacelle improvements for enhanced readiness—all while the NGAS conversation develops in parallel. These aren't competing priorities but rather complementary elements of a coherent modernization strategy that recognizes the Marine Corps cannot afford capability gaps during platform transitions while operating in an increasingly contested global environment.

CONCLUSION: A FORCE IN TRANSFORMATION

The comparison between 2025 and 2026 Marine Aviation Plans reveals a service in the midst of comprehensive, purposeful transformation. While the 2025 plan established foundational concepts and outlined transition pathways, the 2026 plan demonstrates accelerated implementation, organizational maturation, and aggressive pursuit of decision advantage through digital transformation and AI/ML integration.

Several themes emerge consistently across the 2026 plan:

- Distributed Operations as Organizing Principle: Every element of Marine Aviation from platform modernization to sustainment transformation to organizational restructuring is being optimized for distributed operations in contested environments. The elevation of AGS to the seventh function of Marine Aviation, the restructuring of MACGs to establish MAOC, and the aggressive pursuit of digital interoperability all serve the fundamental requirement to operate effectively across vast distances against thinking adversaries.

- Data as the Foundation of Decision Advantage: The 2026 plan positions data-centric warfare not as a future aspiration but as an immediate imperative. The comprehensive framework for transforming aviation sustainment through AI/ML, the detailed roadmap for DI/MANGL implementation, and the integration of Decision-Centric Aviation Operations throughout the force reflect recognition that information superiority enables all other forms of military advantage.

- Balance Between Readiness and Modernization: Throughout the 2026 plan, the tension between maintaining current readiness and accelerating modernization is explicitly acknowledged and deliberately managed. The "Fight Tonight" / "Bridge the Gap" / "Future Fight" framework provides clear guidance for resource allocation, while initiatives like the "26 in 26" safety campaign demonstrate commitment to protecting the force while transforming it.
- Manpower as the Decisive Factor: While the 2026 plan details impressive technological capabilities, it consistently returns to the centrality of Marines, Sailors, and aircrew. The T-E-A-M philosophy, the emphasis on maintenance excellence through Advanced Maintenance Training Academies, and the commitment to strategic talent placement all reflect understanding that technology alone cannot guarantee success.
- Accelerated Timelines and Aggressive Goals: Compared to the 2025 plan, the 2026 version demonstrates increased confidence in execution, reflected in more specific timelines, more ambitious capability targets, and more detailed implementation roadmaps. This reflects organizational learning as transformation progresses from concept to execution.

As Marine Aviation moves forward, the 2026 plan provides a comprehensive blueprint for building the ACE our Nation and Corps requires. By maintaining crisis-response readiness while driving purposeful modernization, Marine Aviation will remain lethal today and prepared for tomorrow's distributed fight.

The success of this transformation will determine not just the future of Marine Aviation, but the Marine Corps' ability to fulfill its role as the Nation's expeditionary force in readiness.

TRANSFORMING MARINE AVIATION SUSTAINMENT: THE AI/ML REVOLUTION IN THE 2026 MARINE AVIATION PLAN

The 2026 Marine Aviation Plan represents a watershed moment in how the Marine Corps approaches aviation readiness. At its core lies a fundamental reconceptualization of sustainment, moving from decades of reactive maintenance practices toward a predictive, data-driven model that leverages artificial intelligence and machine learning to transform how Marine Aviation maintains, supplies, and operates its aircraft.

This transformation directly addresses one of the most persistent challenges facing naval aviation: maintaining high readiness rates while conducting distributed operations across vast geographic areas with limited logistics infrastructure.

The plan articulates the problem with striking clarity: "Marine Corps Aviation remains reactive in maintenance, supply, and operations planning, limiting readiness and reducing the ability to sustain distributed aviation operations and crisis response."

This assessment acknowledges that traditional sustainment approaches built around centralized maintenance facilities, predictable supply chains, and established operational tempos—cannot support the demands of Distributed Aviation Operations (DAO) in contested environments.

When Marine Aviation squadrons must operate from austere, disaggregated sites across the Indo-Pacific or other theaters, the conventional model of flying aircraft back to main operating bases for scheduled maintenance or waiting days for parts to arrive through traditional supply channels becomes operationally prohibitive.

The AI/ML sustainment initiative organizes this transformation into three complementary Lines of Operation: Dynamic Aviation Supply, Predictive Maintenance, and Optimized Operations. Together, these efforts aim to create an integrated sustainment ecosystem where data flows seamlessly between maintenance, supply, and operations systems, where artificial intelligence identifies patterns invisible to human analysts, and where predictive algorithms enable Marines to anticipate and prevent failures before they impact combat readiness.

This represents not merely an incremental improvement in existing processes but a fundamental reimagining of how the Marine Corps sustains its aviation combat power.

LINE OF EFFORT 1: DYNAMIC AVIATION SUPPLY

The first line of effort tackles one of the most vexing challenges in expeditionary aviation: ensuring the right parts are available at the right place and time when operating from distributed, austere locations. Traditional aviation supply packages built around historical consumption data and standardized configurations assume relatively stable operating environments with predictable failure rates and established resupply chains.

These assumptions collapse when squadrons disperse across multiple Forward Arming and Refueling Points (FARPs), expeditionary airfields, and austere sites in contested environments where resupply may be sporadic or impossible for extended periods.

Dynamic Aviation Supply fundamentally rethinks how the Marine Corps structures and employs Supplemental Aviation Spares Support (SASS) packages. Rather than static supply sets based on historical averages, the initiative envisions adaptive packages that respond to actual operational conditions, evolving aircraft configurations, and real-time failure data.

The AI/ML component becomes critical here, machine learning algorithms can analyze vast datasets encompassing aircraft configuration management, operational tempo, environmental conditions, mission profiles, and component failure rates to identify patterns that would be impossible for human logisticians to discern manually.

For example, AI systems can recognize that F-35B aircraft conducting operations in high-temperature, high-humidity maritime environments while carrying external fuel tanks and specific weapons loads experience different component wear patterns than aircraft operating from temperate bases with standard training profiles. The system can then recommend adjusting supply packages for specific deployments to include higher quantities of components statistically more likely to fail under those specific operational conditions, while reducing quantities of parts less likely to be needed.

This precision reduces both the logistics footprint, critical when every pound of supplies must be moved by airlift or ship, and the risk of critical parts shortages that ground aircraft.

The plan's emphasis on capturing "latest configurations and failure rates" addresses another persistent challenge: aviation supply packages often lag behind rapid capability upgrades and configuration changes.

When an aircraft receives new sensors, weapons systems, or defensive countermeasures, the associated supply packages may not immediately reflect the different maintenance requirements and failure modes of the new equipment.

AI/ML systems can continuously analyze maintenance data streams to identify emerging failure trends associated with new configurations and automatically recommend adjustments to supply packages before widespread parts shortages develop.

Beyond individual platform optimization, Dynamic Aviation Supply supports the concept of "highly dynamic, nodal webs of aircraft and support sites" that characterizes Distributed Aviation Operations. In this operating model, aircraft don't return to a single main operating base but may rotate through multiple forward sites, each with varying levels of maintenance capability and parts availability.

AI systems can optimize the distribution of supply packages across this network, ensuring that high-demand items are positioned at nodes with the highest probability of need based on planned operations, aircraft rotations, and historical maintenance trends. This network optimization, balancing readiness requirements against airlift capacity and operational security considerations, represents precisely the type of complex, multi-variable problem where AI/ML excels beyond human cognitive capacity.

The plan explicitly notes goals of "improving deployed aircraft readiness rates while reducing costs and material footprint." These dual objectives often seen as contradictory in traditional logistics become achievable through AI-driven precision. By ensuring the right parts are available where and when needed, the Marine Corps can reduce aircraft downtime (improving readiness) while simultaneously reducing the total quantity of parts that must be procured, stored, and transported (reducing costs and footprint).

Early implementations of similar predictive supply systems in commercial aviation have demonstrated readiness improvements of 15-25% while reducing inventory carrying costs by 20-30%. Achieving comparable results across Marine Aviation would represent a transformational improvement in operational capability.

LINE OF EFFORT 2: PREDICTIVE MAINTENANCE

The second line of effort addresses what may be the most significant cultural and operational shift in the entire AI/ML sustainment initiative: conducting aviation maintenance synchronized with operational tempo and aviation logistics (AVLOG) availability based on data driven probabilities.

Currently, aviation maintenance is categorized as either scheduled or unscheduled. Scheduled maintenance encompasses specific part replacements or inspections based on parameters such as flight hours, calendar days, or events such as the number of takeoffs or landings.

These known parameters allow maintenance and sustainment planners to foresee and plan for impending manpower demands, maintenance time, and supply requirements based on a certain operational tempo. Unscheduled maintenance, >65%-75% of all maintenance actions, are reactionary measures that respond to unanticipated component failures or malfunctions and can severely degrade flight operations and sortie generation.

Employing predictive maintenance concepts, enabled by AI/ML analysis of sensor data, performance parameters, and historical failure patterns, identify impending failures before they occur. This capability facilitates planning for and performing maintenance synchronized with operational aircraft demand, AVLOG availability and location, and manpower available.

Essentially this use of technology turns unscheduled maintenance degraders into foreseen scheduled maintenance actions and supply requirements that can be planned for. This shift profoundly increases aircraft readiness and maintenance efficiency. Further, predictive maintenance bolsters the operational flexibility of a constantly moving force by preemptively locating manpower and supply requirements where they will be needed.

The 2026 plan details several concrete initiatives that establish the foundation for this transformation. The Advanced Maintenance Training Academies for H-1 and V-22 aircraft represent a crucial investment in the human element of predictive maintenance. These programs partner with manufacturers, Bell for the H-1 series and AH-1Z/UH-1Y, and Boeing/Bell for the V-22, to leverage Original Equipment Manufacturer (OEM) instructors and engineering specialists who possess the deepest technical knowledge of these complex platforms.

This partnership approach acknowledges that effectively implementing predictive maintenance requires maintenance personnel to understand not just how to follow technical manual procedures, but why certain failure modes occur, what data signatures indicate developing problems, and how to interpret the outputs of AI-driven diagnostic systems.

The emphasis on "hands-on experience and 3D courseware for critical maintenance tasks" reflects recognition that predictive maintenance demands higher levels of technical expertise than traditional reactive maintenance. When an AI system flags an anomalous vibration signature in an MV-22B proprotor gearbox or identifies unusual temperature trends in an AH-1Z transmission, maintenance personnel must be able to interpret these alerts, conduct focused inspections to verify or rule out specific failure modes, and make informed decisions about whether to remove components for detailed inspection or continue monitoring. This requires a deeper understanding of system engineering, failure modes and effects, and diagnostic techniques than traditional "remove and replace" maintenance.

The mention of establishing "a strong baseline for Maintenance Chiefs while fostering a culture of maintenance excellence" points to the leadership dimension of this transformation. Shifting from reactive to predictive maintenance requires organizational culture change, not just new technology. Maintenance Chiefs must champion the use of data-driven decision-making, encourage their Marines to trust AI-generated alerts even when conventional indicators appear normal, and balance the confidence to defer scheduled maintenance when data shows components are healthy against the judgment to act preemptively when predictive indicators suggest developing problems.

Several specific technologies and programs support this predictive maintenance transformation. The ODSSHI (Osprey Drive System Safety/Health Instrumentation) system exemplifies sensor-based condition monitoring, adding vibration sensors and instrumentation to the MV-22B drive system to enable real-time monitoring of component health.

Rather than replacing proprotor gearboxes on a fixed schedule, maintenance personnel can monitor actual bearing condition, lubrication system health, and structural integrity, replacing components when data indicates degradation rather than at arbitrary hour limits. This approach both improves safety (by identifying developing problems before catastrophic failure) and reduces maintenance burden (by extending component life when condition remains good).

The broader implications extend beyond individual platforms. As predictive maintenance systems mature across the Marine Aviation enterprise, the accumulated data and refined algorithms become increasingly valuable. Machine learning models trained on years of sensor data, maintenance actions, and component failures can identify subtle patterns that predict specific failure modes with high accuracy.

An AI system that has analyzed data from thousands of CH-53K engines across various operating conditions can recognize early indicators of impending turbine blade erosion or bearing wear with far greater reliability than even the most experienced maintenance chief examining a single aircraft.

The plan's emphasis on "fostering a culture of maintenance excellence" acknowledges that predictive maintenance success depends on maintainers trusting and effectively utilizing AI-driven insights.

This requires transparent systems where Marines understand why the AI recommends specific actions, training that builds confidence in data-driven decision-making, and leadership that consistently reinforces the value of predictive approaches. When a predictive system recommends replacing a component that appears to be functioning normally, and that replacement prevents an in-flight failure two weeks later, the resulting success story reinforces cultural adoption across the maintenance community.

LINE OF EFFORT 3: OPTIMIZED OPERATIONS

The third line of effort represents perhaps the most ambitious element of the AI/ML sustainment initiative: fusing data from previously siloed systems to enable comprehensive optimization of flight operations and maintenance.

The plan specifically identifies NALCOMIS (Naval Aviation Logistics Command Management Information System), M-SHARP (Marine Corps Safety and Health Automated Recordkeeping and Reporting Program), and GCSS-MC (Global Combat Support System-Marine Corps) as systems whose data integration will drive optimization. Each of these systems historically operated independently, managed by different communities, and served distinct functions, NALCOMIS for maintenance tracking, M-SHARP for safety and mishap data, and GCSS-MC for supply chain management.

This data fragmentation has long hampered holistic optimization efforts. A squadron operations officer planning the flight schedule sees one picture based on aircraft status codes in NALCOMIS. The maintenance control chief sees overlapping but distinct information about component time remaining and scheduled maintenance requirements. The supply officer monitors parts availability through GCSS-MC. The safety officer analyzes mishap trends in M-SHARP. Each domain optimizes within its silo, but true enterprise optimization requires synthesizing information across all these domains simultaneously.

The AI/ML initiative aims to break down these barriers by creating integrated data platforms where information flows seamlessly between systems and AI algorithms can identify optimization opportunities invisible to humans working within individual functional areas.

Consider a practical example: an AI system analyzing the integrated dataset might recognize that a particular maintenance task on the AH-1Z, when performed by specific maintenance personnel during certain times of day, correlates with slightly elevated mishap risk indicators in subsequent flights. This pattern, involving data from maintenance records (NALCOMIS), personnel qualifications, work schedules, and safety reports (M-SHARP), would be virtually impossible for human analysts to identify across separate databases. The AI can flag this correlation, prompting investigation that might reveal, for instance, that the task involves critical torque specifications more difficult to achieve accurately under high-temperature conditions, leading to procedural adjustments that improve both safety and maintenance quality.

The plan describes developing "a suite of AI-enabled tools to automate and optimize complex, data-intensive tasks of scheduling and managing flight operations and maintenance." Flight scheduling in a Marine Aviation squadron represents an extraordinarily complex optimization problem. Schedulers must balance pilot currency requirements, aircraft maintenance status, training objectives, operational commitments, fuel availability, range time, weather conditions, crew rest requirements, and numerous other variables. Human schedulers rely on experience and judgment to create workable schedules, but AI systems can evaluate millions of potential schedule permutations, identifying options that maximize training value while minimizing maintenance disruption and operational risk.

Maintenance scheduling presents equally complex optimization challenges. When multiple aircraft require different maintenance actions with varying duration, priority, and parts availability, determining the optimal sequence, which aircraft to maintain in which order, which maintenance tasks to perform concurrently versus sequentially, and how to allocate limited maintenance personnel across competing demands, taxes human cognitive capacity.

AI optimization algorithms can simultaneously consider aircraft operational priority, parts availability, maintenance task dependencies, personnel qualifications and availability, facility constraints, and projected operational requirements to generate maintenance schedules that maximize squadron readiness.

The integration of safety data through M-SHARP adds another critical dimension. By analyzing correlations between maintenance actions, operational patterns, and safety incidents, AI systems can identify risk factors that inform both scheduling decisions and procedural improvements. If data reveals that certain mission profiles following specific maintenance actions correlate with elevated precautionary landing rates, the system can recommend additional quality assurance checks or suggest scheduling these mission types with time buffers that allow thorough post-maintenance operational checks.

The plan's vision of "generating safer, more efficient operations that reduce unscheduled downtime and increase aircraft availability" reflects the ultimate objective: using data integration and AI optimization to fundamentally improve how Marine Aviation operates. Safer operations result from AI systems identifying and flagging risk correlations across multiple data sources.

More efficient operations emerge from optimal scheduling that reduces conflicts, minimizes aircraft idle time, and coordinates maintenance with operational tempo. Reduced unscheduled downtime follows from predictive maintenance that addresses developing problems before they cause failures. Increased aircraft availability comes from the cumulative effect of all these improvements, better parts availability, more efficient maintenance, optimized scheduling, and proactive problem prevention.

The challenge of implementing this integrated optimization capability should not be underestimated. It requires not just technology development but also policy changes, procedural updates, and organizational adaptation. Data sharing agreements must be established. System interfaces must be developed. Data quality and standardization issues must be resolved. AI algorithms are only as good as the data they analyze, and years of inconsistent data entry practices across multiple systems present significant challenges. Personnel must be trained to use new tools and trust AI-generated recommendations. Leaders must be willing to make decisions based on algorithmic optimization rather than purely intuitive judgment.

STRATEGIC IMPLICATIONS AND OPERATIONAL IMPACT

The AI/ML sustainment initiative's strategic significance extends well beyond improving maintenance efficiency or reducing parts shortages. This transformation directly enables the distributed operations concept that underlies Marine Aviation's contribution to Expeditionary Advanced Base Operations and stand-in forces operations in contested environments.

When aircraft operate from disaggregated sites with limited maintenance infrastructure and constrained supply chains, the ability to predict failures, optimize limited parts inventories, and efficiently manage maintenance with reduced personnel becomes operationally decisive.

Consider operations across the Indo-Pacific, where Marine Aviation detachments may operate from remote islands or austere expeditionary airfields hundreds of miles from main operating bases. In this environment, an unscheduled maintenance issue that grounds an aircraft might take days or weeks to resolve through traditional supply and maintenance processes, time during which that aircraft's combat power is unavailable and the detachment's operational capability is degraded.

AI-driven predictive maintenance that identifies developing problems before deployment, dynamic supply packages that ensure critical parts are forward-positioned, and optimized scheduling that maximizes available aircraft utilization transform this equation. Aircraft remain mission-capable longer, maintenance issues are resolved faster with forward-positioned parts, and limited maintenance personnel focus efforts where they generate the greatest readiness return.

The initiative also addresses the perennial tension between maintenance and operations tempo. Historically, high operational tempos strain maintenance organizations, leading to deferred maintenance, reduced quality assurance, and eventual readiness degradation. By optimizing maintenance scheduling, improving parts availability, and enabling predictive interventions that prevent major failures, the AI/ML approach allows Marine Aviation to sustain higher operational tempos without degrading readiness. This capability becomes critical during crisis response operations where sustained, high-tempo operations may be required for weeks or months.

The data infrastructure and AI capabilities developed through this initiative also position Marine Aviation rapidly to integrate insights from the broader Joint Force and industry partners. As other services and commercial aviation operators implement similar AI-driven sustainment approaches, the Marine Corps can leverage shared learning, incorporate proven algorithms, and contribute its unique distributed operations insights to the broader community. This collaborative approach accelerates capability development while reducing costs and technical risk.

CONCLUSION: THE PATH FORWARD

The AI/ML sustainment initiative in the 2026 Marine Aviation Plan represents transformational ambition matched with practical implementation focus. By organizing the effort into three clear lines of operation, Dynamic Aviation Supply, Predictive Maintenance, and Optimized Operations, the plan provides a coherent framework for what could otherwise become an unwieldy technology implementation program. Each line of operation addresses specific capability gaps while contributing to the overarching goal of transitioning from reactive to predictive sustainment.

Success will require sustained commitment, significant investment in both technology and training, and willingness to change established processes and organizational cultures. The plan's emphasis on Advanced Maintenance Training Academies, industry partnerships, and cultural transformation alongside technological development reflects realistic understanding that effective AI/ML implementation depends as much on people and processes as on algorithms and data systems.

The ultimate measure of success will be operational impact:

- Can Marine Aviation squadrons sustain distributed operations across contested environments with higher readiness rates and reduced logistics footprint?
- Can predictive maintenance prevent in-flight failures that endanger aircrew and degrade combat capability?
- Can optimized operations and maintenance scheduling generate more available aircraft with existing personnel and resources?

If the AI/ML sustainment initiative achieves these outcomes, it will represent one of the most significant advances in Marine Aviation readiness and operational capability in decades. This would be a transformation as consequential as the technological improvements in the aircraft themselves.

THE 2026 MARINE AVIATION PLAN AND STEEL KNIGHT 25: TURNING PROJECT EAGLE INTO COMBAT PRACTICE

The 2026 Marine Aviation Plan marks a pivot point for Marine Corps aviation, moving from Force Design rhetoric to a structured blueprint for a distributed, data-driven, lethal Aviation Combat Element (ACE). Steel Knight 25, executed by I MEF and 3rd MAW across the Southwest in December 2025, shows what it looks like when that blueprint is stress-tested in a realistic campaign: not with slide decks, but with embassy reinforcements, forward spokes, contested logistics, and kill-web command posts running under real friction.

Taken together, the plan and the exercise reveal a force that is trying to do two things at once: fight tonight with the force you have, and deliberately build the aviation architecture needed for a chaos-management fight against peer adversaries.

FROM FORCE DESIGN TO PROJECT EAGLE AND TO SOUTHERN CALIFORNIA

On paper, the 2026 Marine Aviation Plan is about coherence and pacing. Project Eagle stretches aviation planning across three Future Years Defense Programs, Fight Tonight (2026–2030), Bridge the Gap (2031–2035), and Future Fight (2036–2040), to prevent the service from lurching from crisis to crisis with a FYDP-deep vision. It codifies Distributed Aviation Operations (DAO) as the aviation contribution to Expeditionary Advanced Base Operations (EABO), elevates Aviation Ground Support (AGS) to the seventh function of Marine aviation, and lays out explicit modernization tracks for F-35, MV-22, CH-53K, H-1, MQ-9, collaborative combat aircraft, ground-based air defense, and digital interoperability.

Steel Knight 25 translated those abstractions into a campaign laboratory spread from Camp Pendleton to Mather Airport and Victorville. Rather than a single set-piece battle, it strung together embassy reinforcement, noncombatant evacuation (NEO), distributed fires, and contested logistics into a coherent narrative that forced I MEF to fight as a dispersed, networked MAGTF from the first hours of crisis. The exercise doubled as certification for the regiment deploying as Marine Rotational Force–Darwin, tying experimentation directly to forward deployment in the Indo-Pacific architecture the plan is written to support.

This is where the plan's "Fight Tonight / Bridge / Future Fight" framework becomes operational rather than bureaucratic. Steel Knight was firmly a Fight Tonight event: it used the force actually available in late 2025, F-35Bs and Cs that are still ramping to full Block 4 capability, MV-22s with nacelle improvements only partially fielded, CH-53Es with the first CH-53Ks entering service, H-1 squadrons just achieving full digital interoperability, MQ-9 squadrons early in Pacific integration, and a MACG still transitioning toward the Multifunction Air Operations Center (MAOC) construct.

Yet the scenarios, C2 architectures, and logistics problems it tackled were clearly written against the Future Fight: adversary long-range precision fires, a dense drone and ISR threat, degraded comms, and the requirement to run kill webs across hundreds of miles of contested maritime space.

DISTRIBUTED AVIATION OPERATIONS: FROM CONCEPT TO HUB–SPOKE–NODE

The 2026 plan is explicit: DAO is now the central organizing principle for Marine aviation in support of stand-in forces. DAO is defined not just as dispersal, but as the ability to sustain combat power from multiple expeditionary sites, spokes, FARPs, austere airfields, under contested logistics and electromagnetic conditions, enabled by robust AGS and digital interoperability.

Steel Knight 25 applied that logic with a concreteness planners seldom enjoy. 3rd MAW established aviation spokes at Pendleton and a FARP at Sacramento Mather, roughly 400 nautical miles away, to prove that fuel, ordnance, maintenance and C2 could be pushed well beyond main air stations. MAG-16's MV-22 and CH-53 squadrons used those nodes to move Marines for embassy reinforcement and follow-on operations across a wide battlespace rather than shuttling between a handful of "big bases." MAG-11's F-35Cs operated from Victorville, where MWSS-372 created an expeditionary node capable of rapid hot-refuel and re-arm, validating exactly the kind of AGS-driven DAO the plan elevates.

Retired Lieutenant General Hedelund's observations during Steel Knight underlined the distinction between constructive dispersion and mere scattering of forces. DAO only matters, he argued, if distribution changes how Marines close with the enemy, protect sustainment, and generate effects under pressure.

The exercise made that distinction visible: where hub-spoke-node constructs were tied to clear objectives and time-limited "shelf lives," distribution delivered survivability and tempo; where nodes were too heavy, insufficiently time-bounded, or not well connected to kill webs, dispersion risked becoming an end in itself.

The aviation plan, for its part, tries to institutionalize those Steel Knight lessons. It treats AGS not aircraft as the pacing function for DAO, with specific modernization lines for expeditionary surfacing systems, transportable firefighting and refueling gear, and C-130-movable arresting systems. It also formalizes MAOC as the replacement for stove-piped DASC and TAOC constructs, exactly the integration logic

MACG-38 was working at Steel Knight as it wired MWSS, MASS-3, LAAD, MACS-1, and comms assets into coherent C2 packages for specific vignettes.

KILL WEBS, DECISION-CENTRIC AVIATION, AND THE HUMAN FACTOR

If DAO is about where aviation operates, Decision-Centric Aviation Operations (DCAO) in the plan is about how decisions are made fast enough to matter across that geometry. Project Dynamis and DIMAGTF Agile Network Gateway Link (MANGL) are the aviation plan's answer: make data the primary weapon, integrate sensors, processors, interfaces, and radios across platforms, and use AI/ML to turn raw information into prioritized, actionable cues for commanders and crews.

At 3rd MAW, that language is translated into kill-web problem-solving. The G-6 staff and MACG-38 described their challenge in the simplest terms: separate sensors from shooters, because weapons now outrange onboard sensors, but maintain the speed and precision necessary for contested combat. To do that, they built an "ecosystem" of communications paths, Link-16 for tactical data, military and commercial SATCOM, cellular and fiber when available, and even unclassified networks to move non-sensitive sensor data quickly.

The exercise's most important work here was not technical, though it involved plenty of radios, masts, and satellites. It was cultural. Officers and staff NCOs talked about needing "three ways to win" for every message; about resisting over-centralization as senior headquarters gain more access to forward sensor data and about giving sergeants and lieutenants enough decision authority to execute on commander's intent when a 32-second transmission might get them targeted within minutes.

This is exactly the tension the aviation plan grapples with when it moves from platform roadmaps to digital transformation. It details DI/MANGL as the DI backbone, outlines Extended Tactical Network concepts to extend that backbone to fly-away kits and vehicles, and even proposes AI-enabled tools to fuse data from maintenance and operations systems to optimize schedules and readiness.

But it repeatedly returns to the T-E-A-M framework — Take care of Marines, Execute the basics with brilliance, Attain and maintain readiness, Mitigate risk — as a reminder that human factors still account for nearly 80 percent of mishaps and that digital tools only matter if crews and controllers can actually use them under stress.

Steel Knight made that human dimension visible. In the Wing Operations Combat Center (WOCC), watch officers stressed that the core question was not whether a given system "worked," but whether Marines could integrate it into real mission execution without being overwhelmed or slowed.

At the squadron level, HMLA-267's journey from "forgotten" attack squadron to fully digitally interoperable H-1 unit showed how much mental rewiring is required for aviators to treat information and spectrum as co-equal to weapons.

Major Jonathan Moss's trajectory, from voice-radio Cobra pilot to digital interoperability lead at VMX-1 and then HMLA-267, illustrated the plan's point that transformation is ultimately a cognitive and cultural exercise.

PLATFORMS IN TRANSITION: F-35, MV-22, CH-53K, H-1, AND MQ-9

The aviation plan is, necessarily, platform-heavy. It spells out F-35B/C inventory targets, squadron PAA growth from 10 to 12 jets, TR-3 and Block 4 upgrade paths, and detailed weapon integration timelines for SDB II, AARGM-ER, LRASM, and JASSM. It charts the final glide paths for FA-18 and AV-8B, engineering details of MV-22 nacelle and gearbox improvements, the three-pillar logic for CH-53K deployment (inventory, sustainment, capability), SPINE and PASM for H-1, and the force-structure growth of MQ-9 squadrons.

Steel Knight's contribution was to show what those transitions mean in operational context. MAG-11's F-35Cs were treated less as boutique strike assets and more as long-range, high-endurance sensor and C2 nodes, the "brain" of multi-domain aviation packages that could stretch sensing and networking well beyond the range of legacy platforms, while still delivering kinetic effects when required. Their ability to operate from shorter runways with mobile arresting gear, and to combine partial fuel loads with inflight refueling, mapped directly onto the DAO and AGS constructs in the plan.

MAG-16's CH-53Es and MV-22s, meanwhile, embodied the plan's heavy-lift and tiltrotor storylines. Commanders described routinely flying ranges that outstripped rotary-wing escorts, gambling on tankers that might or might not make it to the rendezvous, and pushing into regimes where the CH-53K's promised lift, stability, and predictive maintenance capabilities will be the difference between concept and reality.

Exercise vignettes highlighted both the potential and the current limitations: on the one hand, rapid employment of spokes and short-duration nodes that fit the S^2D (survivability, sustainability, duration) framework the plan wants to embed in force-insertion thinking; on the other, the continued dependence on USAF C-130s and scarce KC-130Js for long-range logistics, and the immaturity of predictive-maintenance and 3D-printing ecosystems to support truly lean detachments.

For H-1s and MQ-9s, the alignment was striking. The plan's SPINE, PASM, and DI lines for the H-1 community are written against a vision of Cobras and Hueys as digital, multi-role nodes in kill webs rather than just close-support trucks. HMLA-267 at Steel Knight was that vision in its first operational iteration: a fully DI "unit of employment" using Link-16 to plug into the same digital ecosystem as F-35s, MQ-9s, E-2s, Growlers and ground C2 nodes, compressing the find-fix-track-target-engage-assess cycle and serving as C2 extensions for stand-in forces.

Similarly, the plan's expansion of MQ-9A squadrons and capability sets, Sky Tower relay, persistent LEO links, electronic support, DAAS kits, is written to make VMUs central to distributed C2 and maritime domain awareness.

Steel Knight's integration of unmanned surface vessels into the air picture via unclassified networks, and its emphasis on making any sensor feed any shooter, showed how 3rd MAW and MACG-38 are already thinking in those terms even as MQ-9 fielding and integration continues.

AIR DEFENSE, C2, AND THE MOVE FROM BASES TO HUBS

The plan dedicates significant space to ground-based air defense (GBAD), with Marine Air Defense Integrated System (MADIS) scaling to 190 systems and Medium Range Intercept Capability (MRIC) achieving IOC in FY26, and to the restructuring of MACGs around MAOC and modernized air traffic control. It also emphasizes G/ATOR as the multi-mission radar tying MRIC, air control, and counter-battery functions together, and calls for reactivating a reserve LAAD battalion to close long-standing gaps.

Steel Knight highlighted why those investments matter. MACG-38 commander Colonel O'Connell described his group not as a supporting element but as the "dial" that configures aviation for distributed operations: combining MWSS, MASS-3, MACS-1, 3d LAAD, and the comms squadron into modular packages that create specific effects in time and space.

Air defense, which spent the Iraq and Afghanistan years on the margins or serving as provisional infantry, is now central: MADIS and future MRIC batteries must protect nodes and hubs against a saturated UAS and missile threat, and must be integrated into the same fires and maneuver logic as artillery and aviation.*

Lieutenant Colonel Astacio's account of MACS-1's evolution

---

* MADIS is the U.S. Marine Corps' new short-range, ground-based air defense system, built around Joint Light Tactical Vehicles (JLTVs) and designed to detect, track, and engage unmanned aerial systems (UAS), cruise missiles, artillery, rockets, mortars, and manned aircraft at low altitudes. It integrates radar, Stinger missiles, a 30 mm cannon, and command-and-control elements into a mobile, networked "kill chain" that can move with Marine maneuver units. MRIC refers to the Marine Corps' planned medium-range surface-to-air missile batteries, based on the Israeli Iron Dome system and using Tamir interceptors. MRIC batteries are designed to defend fixed and semi-fixed sites (nodes and hubs) against cruise missiles, rockets, artillery, mortars, and drones, filling a medium-range air-defense gap left after the retirement of the HAWK system. They are intended to be interoperable with theater air-defense architectures and integrated into broader fires and maneuver schemes alongside artillery and aviation.

showed the hardware side of that shift. In a decade, the unit has gone from "big box" radars set up as enduring fixtures to the G/ATOR, a single radar replacing five legacy systems, transportable by C-130 or potentially CH-53K, and designed to feed expeditionary command posts that move, operate under EMCON, and distribute sensing via both active and passive systems. Steel Knight forced MACS-1 and MAOC-like elements to actually move C2 nodes, manage bandwidth and risk, and maintain decision-quality information without the comfort of a fixed tower at a secure base.

The plan codifies that direction by formally transitioning to MAOC, restructuring active and reserve MACGs, modernizing expeditionary ATC, and aligning MANGL/MAGTAB integration with forward air sites and FARPs. Steel Knight, in turn, exposes where that transition is incomplete: parts availability for G/ATOR, the deployment lift burden for MAOC-scale packages, remaining seams between F-35, G/ATOR, and GBAD data, and the still-evolving understanding of how much C2 authority and functionality must reside at forward hubs versus rear wings and MEF headquarters.

CHAOS MANAGEMENT
AS THE LINKING THEORY

Both the 2026 aviation plan and Steel Knight 25 are wrestling with the same deeper shift: from crisis management to chaos management. The plan hints at this in its emphasis on decision-centric operations, Project Dynamis, and a Fight Tonight / Future Fight continuum rather than neat phase lines between "competition," "crisis," and "war." Steel Knight makes it explicit.

Modern operations unfold within persistent complexity rather than discrete, bounded crises. Adversaries test and probe in the gray zone;. Satellite comms can be disrupted. Centralized command posts can be targeted. Cyber and information operations can create enduring ambiguity. In that environment, a Marine aviation enterprise optimized for surge from secure bases is insufficient. The force must be designed to live in chaos: sustainable presence rather than episodic surge, distributed decision authority rather than centralized control, resilient nodes rather than exquisite but brittle hubs.

Project Eagle, DAO, DCAO, DIMAGTF, AGS elevation, MAOC, MQ-9 growth, GBAD expansion, all of those plan artifacts make sense only when read against that chaos-management backdrop. Steel Knight 25 is where those artifacts get their first full-up operational rehearsal. Marines discover that tanker density and heavy-lift short-falls will limit what DAO can really do, that predictive maintenance and AI-driven sustainment tools must mature if small nodes are to remain viable, that mission command has to be more than a slogan if distributed Marines are to exploit fleeting targets without over-the-horizon micromanagement; and that command posts must be as mobile and signature-managed as the aircraft they control.

In that sense, the relationship between the 2026 Marine Aviation Plan and Steel Knight 25 is symbiotic. The plan gives structure, resources, and sequencing to the transformation. Steel Knight supplies the operational truth serum, revealing where that structure is well-founded and where it must be reworked.

Together, they are writing, not just in concept, but in practice, the architecture of a Marine Aviation Combat Element designed not for tidy crises, but for the enduring, contested chaos of 21st-century military operations..

CHAPTER 4

FORCE DESIGN MEETS REALITY: WHAT STEEL KNIGHT 25 REVEALS ABOUT THE MARINE CORPS' STRATEGIC RESET

The Marine Corps stands at a pivotal moment in its transformation. The October 2025 Force Design Update and the Steel Knight 25 exercise held in December 2025 together tell a story of strategic adaptation, one where ambitious concepts meet the harsh judgment of operational reality, and where institutional humility trumps doctrinal rigidity.

The original Expeditionary Advanced Base Operations (EABO) concept envisioned Marine units island-hopping across the Pacific, armed with anti-ship missiles, creating dilemmas for adversary fleets through distributed lethality. It was an appealing vision, rooted in Marine Corps amphibious heritage while addressing modern anti-access challenges.

But rather than defending an elegant concept that field testing revealed as problematic, Marine Corps leadership adjusted based on evidence. The shift acknowledges logistical constraints, survivability challenges, and the comparative advantage Marines actually possess: the ability to persist forward with advanced sensors and communications, serving as the eyes and ears for long-range precision fires launched from platforms with greater reach and magazine depth.

THE THIRD MLR DECISION: PRESERVING FLEXIBILITY

Equally significant is the decision to retain 4th Marine Regiment as a reinforced infantry regiment rather than converting it into a third Marine Littoral Regiment. This reverses a core element of the original Force Design plan and directly addresses criticism that the transformation cut too deeply into conventional infantry capacity.

The retention of 4th Marines as a conventional infantry regiment represents a hedge against strategic uncertainty. While two MLRs provide specialized capabilities for contested littoral operations, maintaining traditional infantry preserves capacity for contingencies outside the First Island Chain scenario. Recent operations near Venezuela, highlighted by critics, underscore this need for versatile forces not optimized exclusively for one operational problem.

Importantly, 4th Marines isn't simply being left behind. The Force Design Update specifies that "previously programmed MLR-associated equipment and personnel" will flow to III MEF to be task-organized as commanders require. This creates flexibility without abandoning the specialized capabilities (NMESIS, G/ATOR, MADIS) that define the MLR concept.

STEEL KNIGHT 25: THEORY MEETS FRICTION

Steel Knight 25 provided a critical testing ground for these evolving concepts. Large-scale exercises serve purposes beyond training. They reveal friction points between doctrine and reality that no amount of staff planning can anticipate.

The exercise emphasized several Force Design priorities:

- Digital Interoperability: Connecting sensors to shooters across domains remains central to the "kill web" concept. Steel Knight tested whether distributed Marine units could actually detect, track, and pass targeting data to joint fires platforms under realistic conditions. The integration of Live-Virtual-Constructive (LVC) training environments through Project Tripoli allowed exercise planners to simulate threats and capabilities not physically present, creating complexity that pure live-force exercises cannot achieve.

- Distributed Operations: Operating from multiple dispersed locations tests command and control resilience, logistics sustainment, and tactical coordination in ways that concentrated operations do not. The electromagnetic environment in Steel Knight likely included jamming, communications degradation, and cyber effects, forcing units to operate with the degraded connectivity they'll face in actual conflict.

- Joint Integration: If Marines are shifting toward an enabling role for joint kill webs, exercises must validate that integration works. Steel Knight tested whether Marine sensors and targeting data could actually flow into Navy, Air Force, and Army fires systems with sufficient speed and accuracy to be operationally relevant.

The exercise also exposed continuing challenges. Systems like NMESIS, while fielded to 3rd MLR, continue facing "delays and setbacks" in reaching full battery strength. Ground/Air Task Oriented Radar (G/ATOR) systems, fielded to nearly 60% of planned capacity by FY25, represent significant progress, but also highlight that key capabilities remain incomplete. These gaps matter profoundly in contested environments where capability shortfalls translate directly into operational limitations.

THE INFANTRY BATTALION QUESTION

Both the Force Design Update and likely feedback from exercises like Steel Knight have driven significant changes to infantry battalion structure. The document notes "two in-stride adjustments" based on FMF feedback and observations from the Campaign of Learning, with more changes expected from ongoing Infantry Battalion Experimentation (IBX).

The return to the 13-Marine rifle squad with three fire teams led by a school-trained sergeant represents one such adjustment. The addition of a "precision fires Marine to operate our small lethal drones" in each squad reflects battlefield realities from Ukraine and elsewhere, small unmanned systems have become essential to tactical operations, not specialized capabilities for select units.

The establishment of a "Fires and Reconnaissance company" integrating manned and unmanned ISR with organic fires capabilities suggests another lesson learned: that sensing and striking must be organizationally integrated at lower echelons rather than coordinated across separate units. This reduces coordination friction and accelerates the sensor-to-shooter timeline.

Steel Knight likely revealed specific shortfalls that drove these adjustments, perhaps squads lacking organic ISR struggled with situational awareness, or perhaps fires coordination proved too cumbersome across organizational boundaries. The document's carefully neutral language ("in-stride adjustments") masks what probably involved difficult operational lessons about structures that seemed sound in planning but proved inadequate under exercise pressure.

THE QUIET WALK-BACKS

Perhaps most telling are capabilities the Marine Corps is quietly restoring after earlier divestment. The Force Design Update notes that all three Marine Expeditionary Force elements are "exploring gap crossing and obstacle breaching capabilities to ensure our forces can close with and destroy the enemy."

Bridging capabilities, including armored vehicle launched bridges, were divested alongside tanks in the original Force Design plan as the service worked to become lighter and redirect investments toward anti-ship weapons. The reversal acknowledges that the Joint Force cannot meet all Marine Corps needs in this area and that the Army's breaching systems are "probably not expeditionary enough for Corps requirements."

This illustrates how exercises expose second and third-order effects of force structure decisions. Divesting tanks seemed logical if the focus was anti-ship operations from island positions. But mobility across contested littoral terrain where adversaries will employ obstacles, mines, and complex defenses requires breaching capabilities that got divested along with heavy armor. Presumably, Steel Knight or similar exercises revealed scenarios where Marine units couldn't maneuver because they lacked tools to overcome obstacles that any competent adversary would employ.

THE CAMPAIGN OF LEARNING METHODOLOGY

What distinguishes the current approach from previous transformation efforts is the explicit embrace of experimentation and adaptation. The Force Design Update emphasizes that modernization "remains a continuous Campaign of Learning and adaptation" and not a fixed endpoint.

To date, that campaign has included more than 70 studies, 33 experiments, 42 war games, and 45 planning teams. This represents substantial investment in understanding what works before committing fully to new structures and capabilities. Steel Knight 25 fits within this broader framework, one large-scale iteration testing multiple concepts simultaneously.

The integration of exercises like Steel Knight with the Marine Corps Lessons Learned Program (MCLLP) and Training and Education Command's Trend Reversal and Reinforcement Process (TRRP) creates mechanisms for converting operational experience into institutional learning. The development of the Analytic Master Plan-Marine Corps (AMP-MC) aims to formalize this process, ensuring analysis drives decisions rather than advocacy or institutional inertia.

IMPLICATIONS AND PATH FORWARD

The relationship between Force Design October 2025 and Steel Knight 25 illustrates pragmatic adaptation in action. Rather than rigidly implementing a predetermined vision, the Marine Corps is using exercises to test concepts, identify problems, and adjust accordingly. This iterative approach reflects lessons from observing rapid battlefield evolution in Ukraine, where forces that adapted quickly gained advantage over those wedded to doctrine.

Several implications emerge:

- Continued Evolution is Inevitable: If the shift from shooter to sensor took five years to fully recognize, other adjustments will emerge as new capabilities field and exercises grow more sophisticated. The Marine Corps should expect its 2030 force to look different from current plans.
- Joint Integration Matters More Than Independence: The sensor emphasis acknowledges that Marines operate as part of a larger joint force, not as an independent service. This has implications for acquisition, training, and operational planning.
- Geography Still Constrains: While the decision to retain 4th Marines provides flexibility, the bulk of specialized capabilities remain in III MEF. This suggests the Indo-Pacific focus remains dominant despite rhetoric about global responsiveness.
- Industrial Base Challenges Persist: The Force Design Update notes continuing struggles to establish supply chains for new systems like NMESIS. Fielding delays directly impact operational readiness and limit what exercises can actually test.

The Marine Corps' willingness to adjust Force Design based on exercise results deserves recognition. Too often, military services defend transformation plans against evidence of problems, allowing institutional investment and ego to override operational reality. The decisions to retain 4th Marines, restore breaching capabilities, and shift emphasis from direct fires to targeting support all suggest an institution capable of learning.

The Marine Corps' transformation isn't complete, and shouldn't be.

What matters is maintaining the intellectual humility and operational pragmatism that allowed Force Design to evolve from its 2020 vision to its 2025 reality. Steel Knight 25 represents one iteration in a continuous process of learning, adapting, and refining the force to meet challenges that remain partially unknowable until tested in the crucible of realistic operations.

CHAPTER 5

AN OVERVIEW ON THE TRANSITION

THE KILL WEB: A NEW ERA OF INTEGRATED MILITARY AVIATION

The modern operational environment has undergone a tectonic shift. We are no longer in a world of "Crisis Management", the episodic response to discrete, predictable events like embassy reinforcements or humanitarian relief.

Instead, the Joint Force must now master Chaos Management. This is a permanent state of multi-domain competition where the boundary between peace and war is blurred, and threats emerge simultaneously across geographic and functional domains.

In this era, the commander must assume that every domain is contested and that the luxury of a secure, linear build-up of force no longer exists. Success requires a transition from platform-centric responses to a network-enabled, effects-based mindset.

Managing this chaos requires a move away from the brittle structures of the past toward a resilient, non-linear architecture known as the Kill Web.

Anatomy of the Kill Web:
Moving Beyond the Chain

For decades, military doctrine relied on the Kill Chain, a linear, step-by-step sequence: find, fix, track, target, engage, and assess. In a chain, any single broken link, a jammed radio or a destroyed sensor, fails the entire mission. The Kill Web replaces this fragility with a dynamic, multi-nodal network where every sensor, decision-maker, and shooter is interconnected through multiple redundant pathways.

- Resilience through Redundancy: In a web, any vetted sensor (e.g., a drone or forward observer) can cue any appropriate shooter (e.g., a ship or aircraft) via several different digital paths. So What? This eliminates "single points of failure." If an adversary destroys a primary command node, the force remains lethal and connected through the remaining mesh.
- Dynamic Connection: The web enables "any-to-any" connectivity, allowing information to bypass traditional hierarchies. So What? It creates an adaptable force that can reconfigure in real-time. A pilot can instantly pass a target to a naval vessel hundreds of miles away that possesses a more optimal firing solution.
- Sensor-to-Shooter Speed: By leveraging digital interoperability and AI-assisted tools, the web compresses the time between detection and action. So What? In modern combat, a 32-second radio transmission is an invitation to a kinetic strike. The Kill Web allows data to move at machine speed, ensuring the Joint Force acts before the adversary can react.

While the Kill Web is a conceptual framework, its effectiveness is realized through the physical aircraft that serve as its critical "nodes."

The Aircraft as a Network Node

In the Kill Web model, military aircraft are no longer viewed as isolated weapons platforms defined by their individual "kill counts." They are information hubs, network nodes, that gather, process, and distribute data to enable the entire Joint Force.

- F-35 (The Sensor-Shooter-Node): Beyond its role as a stealth fighter, the F-35 is a penetrating sensor and "airborne brain." It operates where legacy platforms cannot, using advanced Electronic Warfare (EW) capabilities and sensors to discover targets and distribute high-fidelity data across the web, acting as a catalyst for joint lethality.
- H-1 Helicopter (The Digital Warrior): Once seen as legacy shooters, the H-1 community (specifically HMLA-267 as the proof-of-concept squadron) has led a digital revolution. These aircraft now function as digital gateways, bridging the gap between legacy voice systems and modern high-speed data links to ensure forward ground units remain connected to the web.
- MQ-9 (Persistent Eyes): Through high-endurance sensing, the MQ-9 provides the "fabric of decision advantage." By maintaining custody of a target for over 20 hours, it allows commanders to anticipate adversary movements and shape the battlespace days before a kinetic engagement occurs.

To remain effective, these nodes must operate within a mobile framework that prevents the adversary from fixing their location.

DISTRIBUTED OPERATIONS: THE ART OF STAYING MOBILE

The Impact Force survives and wins through Distributed Aviation Operations (DAO). This strategy spreading assets across a Hub-Spoke-Node taxonomy to complicate enemy targeting and ensure persistence:

- Hubs: Large bases with robust infrastructure and sustainment.
- Spokes: Austere, temporary locations for short-term mission support.
- Nodes: Highly transient, mobile spots existing for a narrow mission window.
- The S2D Framework and Timestamped Nodes Sustainability in a contested environment is governed by the S2D framework: Survivability, Sustainability, and Duration. Every forward node must be viewed as having an inherent shelf life—they are "Timestamped Nodes."
- Signature Management: Every transmission and movement creates an electronic or physical signature.
- Rapid Retribution: Because an adversary can strike within minutes of a 32-second transmission, staying stationary is a death sentence.
- Execution and Evacuation: The mindset must shift from "defending a base" to "executing an effect and immediately evacuating." Movement is a prerequisite for survival and subsequent decision-making.

Assault Support: The Mobility Backbone

This distributed posture is enabled by the "Assault Support" community, which serves as the physical connector of the web:

- MV-22 Osprey: Uses its unique speed and range to insert and extract Marines from non-traditional zones, making the force's location unpredictable.
- CH-53K King Stallion: The heavy-lift workhorse capable of establishing a functioning node—carrying fuel, ammunition, and power—in a single flight. However, commanders must recognize that logistics is not "magic"; even the CH-53K requires sufficient mass to overcome the "voodoo" expectations of sustaining a distributed force at scale.

Sensing and moving, however, are only relevant if they lead to superior decision-making

Decision Advantage: The Human Factor

The ultimate objective of the Kill Web is Decision-Centric Aviation Operations (DCAO). In this paradigm, information is the primary weapon. The goal is to generate "Decision Advantage"—the ability to sense, process, and act faster and more accurately than the enemy.

The 78.8 Percent Problem and Cognitive Load Data from the source indicates that nearly 80% (78.8%) of aviation mishaps involve human factors. As the Kill Web increases the volume of data, the cognitive load on pilots and controllers increases exponentially. DCAO uses AI-enabled tools to filter this noise, allowing the human to focus on the most critical tactical choices.

The Delegation Dilemma There is a fundamental tension between Centralized Oversight and Mission Command.

Technical architectures now give junior captains and sergeants at forward nodes access to the same sensor data as generals.

The Dilemma: If a forward node must wait for permission from a distant headquarters to act on a fleeting target, the technical speed of the Kill Web is neutralized. Success requires a culture of trust where decision authority is delegated to the "edge" of the network.

Summary: The "Impact Force" Mindset

The transition to a Kill Web model marks the emergence of the Impact Force. This is a critical evolution from the earlier "Inside Force" concept. While an Inside Force focuses on geographic positioning to survive within a threat ring, the Impact Force focuses on operational effect generation. As General Haar noted, the goal is for the forward-deployed force to act as the "JTAC for the Joint Force"—sensing and enabling the entire team to win.

- From Position to Effect: A platform's value is no longer its individual kill count but its "operational impact"—how much it increases the lethality and decision speed of the Joint team.

- Digital Interoperability as a Priority: The most lethal "weapon" on a modern aircraft is its ability to talk to other systems. Digital connectivity is the prerequisite for the Kill Web.
- Logistics is the Limiting Factor: You cannot execute distributed operations with "magic." Contested logistics—the ability to fuel, fix, and move under fire—is the foundation of the Impact Force.
- Culture of Trust: Technology alone will fail if command remains centralized. Commanders must embrace Mission Command, trusting junior leaders at the edge to make high-stakes decisions based on the data they see.

STRATEGIC TRANSITION ANALYSIS: FROM CRISIS RESPONSE TO JOINT IMPACT FORCE

The United States Marine Corps is currently executing a fundamental strategic pivot, necessitated by the terminal obsolescence of the post-Cold War crisis-management model. For decades, the service operated under an assumption of regional stability occasionally punctuated by discrete, time-bounded emergencies. However, the rise of peer-competitor gray-zone operations has effectively eroded the binary boundaries between peace and war, rendering the traditional, self-contained Marine Air-Ground Task Force (MAGTF) a systemic liability within a contested Weapons Engagement Zone (WEZ). In an era of pervasive surveillance and long-range precision fires, the legacy "Force Closure" model—relying on methodical, permissive buildup—invites catastrophic targeting rather than operational readiness.

This transition requires a sophisticated shift from "Crisis Manage-ment" to "Chaos Management." Crisis management relies on rehearsed, linear playbooks and predictable escalation ladders. Conversely, chaos management assumes a permanent state of multi-domain threats, where compressed decision cycles and signature-management requirements outpace traditional hierarchical structures. The Marine Corps can no longer function as a standalone responder on the periphery; it must integrate into a broader architecture to survive. The core thesis of the "Impact Force" concept is its role not as an isolated striker, but as a decisive joint-force enabler. By persisting inside the adversary's threat ring, the Impact Force provides the sensing, targeting, and command-and-control (C2) layer that allows long-range joint assets—Navy surface combatants and Air Force bombers—to function with disproportionate lethality. This evolution redefines the Marine Corps' primary value proposition from organic magazine depth to system-wide decision advantage, moving from geographic positioning to operational effect generation.

The strategic refinement of the Marine Corps' role involves transitioning from the static geographic positioning associated with early "Inside Force" concepts to the intentional generation of operational effects. While the initial thinking focused on the physical survival of units within a threat ring to deny sea control, the 2025-2026 evolution emphasizes the Impact Force as a functional connector within a broader kill web.

As Brigadier General Christopher Haar, Assistant Deputy Commandant for CD&I, noted, the original focus was "all about killing ships for the stand-in force." The mature "Impact Force" model, however, leverages a "Joint JTAC" metaphor. Just as a Joint Terminal Attack Controller does not drop the bomb but creates the essential conditions for success, the Marine Corps now functions as the "Joint JTAC for the joint force."

This shift from "ship-killing" to "joint sensing and targeting" maximizes the utility of Marine units within the WEZ; by providing track-quality data to shooters with deeper magazines, small units generate strategic impact far beyond their organic weight. This conceptual framework was subjected to rigorous validation during the Steel Knight 25 exercise.

Operational Validation: Steel Knight 25 as a Campaign Laboratory

Steel Knight 25 represented a watershed moment for the Marine Corps, shifting from a standard readiness check to a full-scale "campaign laboratory." The exercise functioned as a continuous rehearsal of distributed entry operations, linking complex problems, embassy reinforcement, noncombatant evacuations, and distributed fires, into a single, high-fidelity operation across the American Southwest. It was the first comprehensive stress test of the Marine Corps fighting as a networked Impact Force under chaos management conditions.

Key lessons from "Distributed Entry Operations" confirm that the traditional model of "force closure" is obsolete in the Indo-Pacific. Steel Knight 25 assumed zero sanctuary and no permissive buildup window, testing I MEF's ability to fight as a networked stand-in force from H-Hour.

This required immediate dispersal and the integration of fires under constant surveillance. A critical finding was the "Heavy Node" problem: current command and logistics nodes possess large physical and electromagnetic signatures, making them easily targetable.

Technical data from the field suggests that even a 32-second radio transmission can invite a lethal enemy response within two minutes. To counter this, the exercise validated the "Timestamped Nodes" concept proposed by LtGen (Ret.) Robert Hedelund.

Rather than treating nodes as permanent assets, they are temporary constructs with inherent "shelf lives." Under this logic, a node is assumed "out" by a predetermined time unless instructed otherwise, forcing a cycle of constant movement where establishment and displacement are synchronized to outpace adversary targeting cycles.

Furthermore, the exercise served as the certification venue for the Marine Rotational Force–Darwin (MRF-D). By tying theoretical experimentation to a forward-deployed unit, the Corps ensured that Impact Force concepts remained grounded in operational discipline and the realities of the Indo-Pacific theater. This operational validation highlights that the primary technological multipliers of the Impact Force are found in advanced aviation and digital interoperability.

Aviation as the Connective Tissue: Project Eagle and the 2026 Plan

The Aviation Combat Element (ACE) is the primary enabler of the Impact Force, serving as the connective tissue that binds distributed units into a cohesive whole. This transformation is managed through "Project Eagle," a three-horizon framework designed to transition aviation from platform-centric to data-centric operations (DCAO):

- Fight Tonight (2026-2030): Immediate readiness and the integration of current inventory into the joint kill web.
- Bridge the Gap (2031-2035): Modernization of existing assets alongside the incorporation of autonomous collaborative combat aircraft.
- Future Fight (2036-2040): A fully realized, data-centric aviation enterprise optimized for chaos management.

Specific platforms, the F-35, MQ-9, and H-1, have transitioned from isolated shooters to "Digital Gateways" and "Airborne Nodes." An F-35 sortie is no longer judged solely by ordnance dropped but by the track-quality data it fuses and shares. The H-1 has evolved into a rotary-wing C2 platform, bridging the digital gap between high-altitude sensors and ground-level stand-in forces. Crucially, the 2026 Marine Aviation Plan has elevated "Aviation Ground Support" (AGS) to the seventh function of Marine Aviation.

This is not a clerical change; it is a logistical revolution. The ability to generate fuel, power, and maintenance from austere, shifting locations (S^2D logic: survivability, sustainability, duration) is the decisive precondition for an Impact Force. Without AGS-enabled "hubs, spokes, and nodes," the aviation enterprise loses its relevance in a contested environment. This logistical foundation is what supports the technical architecture of the broader Kill Web.

BUILDING THE KILL WEB: C2 AND DIGITAL INTEROPERABILITY (DI)

The Impact Force model necessitates a shift from linear "Kill Chains" to resilient, multi-path "Kill Webs." A kill chain is fragile. A single broken link renders the system inert. A kill web allows any vetted sensor to cue any appropriate shooter through redundant pathways. To survive the 32-second transmission risk in a contested electromagnetic environment, the "Three Ways to Win" communications strategy provides necessary redundancy through:

- Link-16: Standardized tactical data exchange for sensor-to-shooter integration.
- Military SATCOM: Protected, beyond-line-of-sight connectivity.
- Commercial LEO (e.g., Starlink): High-capacity bandwidth for rapid data movement.
- Unclassified Networks: Leveraged to rapidly move non-sensitive sensor data when speed is the primary survival metric.

The technical backbone of this web is the Multifunction Air Operations Center (MAOC) and the Marine Air Defense Integrated System (MADIS). The MAOC provides the expeditionary C2 hub, while MADIS provides the organic "shield" against cruise missiles and UAS.

Together, these systems allow small, forward detachments to persist inside a WEZ and feed the joint network. These technical capabilities are further refined by the strategic calibrations found in the 2025 Force Design Update.

Strategic Recalibrations: The 2025 Force Design Update

The October 2025 Force Design Update reflects a "learning organization" capable of evidence-based iteration. The update incorporates empirical data from campaign laboratories like Steel Knight to adjust force structure. Key course corrections include the decision to retain the 4th Marine Regiment as a reinforced infantry regiment, rather than converting it to a Marine Littoral Regiment (MLR).

This ensures the Corps balances specialized littoral operations with traditional global responsiveness. Additionally, the update restores bridging and breaching capabilities previously divested, acknowledging that even a sensor-centric force must maneuver through contested terrain.

The Marine Littoral Regiment has evolved from a purely sea-denial asset to a sensor-centric enabler for joint kill webs. This shift underscores the Corps' commitment to providing the "forward eyes" of the joint force, prioritizing the movement of information over the movement of mass. These adjustments ensure the force remains lethal and flexible, though significant human and logistical challenges remain.

The Human & Logistical Constraint: Culture and Sustainability

The strategic risk facing the Marine Corps is that technical architecture will outpace command culture and logistics capacity. The "Delegation Dilemma" remains the primary cultural friction point: for Mission Command to be effective, senior leaders must resist the urge to centralize control simply because they have access to the same data as the forward commander. Junior leaders at distributed nodes must be empowered with the authority to act on sensor data immediately, as the 32-second transmission rule leaves no room for centralized approval cycles.

Logistics is the Limiting Factor of the Impact Force. Distributed operations require a radical reinventing of the supply chain. Critical gaps include:

- Lift Capacity: The CH-53K King Stallion is a transformational multiplier capable of assuming missions once reserved for the KC-130J, but it must be acquired in sufficient numbers to meet theater requirements.

- Fuel and Power Distribution: The difficulty of sustaining moving nodes in austere environments cannot be underestimated.
- Reliance on Joint Enablers: Current operations rely heavily on Air Force C-130s. As practitioners have noted, "you cannot do this operation with magic" or the "voodoo" of assuming joint lift will always be available is a strategic vulnerability.

The Marine Corps must trade the familiar, scripted risks of crisis management for the harder-to-predict risks of chaos management. This path requires a culture of radical delegation and a reinvented logistical backbone. By embracing its role as the connective tissue and sensing heart of the Joint Force, the Marine Corps is positioning itself as the decisive impact force in an age of chaos.

STUDY GUIDE: BUILDING THE IMPACT FORCE

This study guide provides a comprehensive review of the strategic and operational transformation of the United States Marine Corps as it transitions from a traditional crisis-response model to a modern "impact force" optimized for chaos management.

Part I: Short-Answer Quiz

Instructions: Answer the following ten questions in 2–3 sentences, drawing specifically from the provided source context.

- How does the document define the difference between a crisis management force and a chaos management force?
- What is the fundamental distinction between an "inside force" and an "impact force" as described in the text?
- What was the primary purpose of Exercise Steel Knight 25 within the Marine Corps' "campaign of learning"?

- How does a "kill web" improve upon the traditional "kill chain" model?
- What is the significance of the Marine Corps' pivot from being "ship killers" to acting as the "JTAC for the joint force"?
- Explain the "three-horizon" construct of Project Eagle.
- What is the "seventh function" of Marine aviation, and why was it elevated to this status?
- Describe the concept of "timestamped nodes" as advocated by Lieutenant General (Retired) Robert Hedelund.
- How has the role of the F-35 evolved within the impact force framework?
- What are the "three ways to win" regarding digital interoperability and communications?

PART II: ANSWER KEY

How does the document define the difference between a crisis management force and a chaos management force? A crisis management force is optimized for discrete contingencies and linear escalation using established playbooks in a relatively stable environment. In contrast, a chaos management force is designed for persistent competition and non-linear, multi-domain threats where the boundary between peace and war is blurred.

What is the fundamental distinction between an "inside force" and an "impact force" as described in the text? An inside force focuses on geographic positioning and organic lethality to survive and strike within an adversary's weapons engagement zone. An impact force generates disproportionate effects by acting as a sensor, targeting, and command-and-control enabler that amplifies the lethality of the entire joint kill web.

What was the primary purpose of Exercise Steel Knight 25 within the Marine Corps' "campaign of learning"? Steel Knight 25 served as a "campaign laboratory" to test the Marine Corps' ability to operate as a distributed, digitally integrated impact force under realistic chaos conditions. It moved concepts from "slideware" to practice by linking complex problems like embassy reinforcement, distributed fires, and contested logistics.

How does a "kill web" improve upon the traditional "kill chain" model? A traditional kill chain is a linear sequence that collapses if any single link is broken. A kill web is a dynamic network where any vetted sensor can cue any appropriate shooter through multiple redundant pathways, ensuring resilience even when specific nodes are degraded.

What is the significance of the Marine Corps' pivot from being "ship killers" to acting as the "JTAC for the joint force"? This shift acknowledges that Marine Corps value is maximized when forward units use their sensors and "eyes" to enable deeper joint magazines rather than focusing solely on their own organic weapons. By acting as a "Joint Terminal Attack Controller" (JTAC) for the entire joint force, the Corps provides the necessary identification and timing to close kill webs with high-end precision.

Explain the "three-horizon" construct of Project Eagle. Project Eagle stretches planning across three Future Years Defense Programs: FYDP-1 (2026–2030) focuses on "Fight Tonight," FYDP-2 (2031–2035) aims to "Bridge the Gap," and FYDP-3 (2036–2040) prepares for the "Future Fight." This structure allows for the deliberate sequencing of capability changes to ensure distributed operations grow more robust over time.

What is the "seventh function" of Marine aviation, and why was it elevated to this status? Aviation Ground Support (AGS) was elevated to the seventh function of Marine aviation to recognize that distributed operations are unsustainable without the ability to generate fuel, power, and maintenance at austere, shifting locations. AGS provides the backbone required to turn Distributed Aviation Operations (DAO) from a concept into an operational reality.

Describe the concept of "timestamped nodes" as advocated by Lieutenant General (Retired) Robert Hedelund. LtGen Hedelund argues that distributed nodes should have inherent "shelf lives" measured in hours or days to avoid detection and targeting. Under this concept, a node is assumed to be extracted at a specific time unless told otherwise, forcing constant movement to ensure survival in contested environments.

How has the role of the F-35 evolved within the impact force framework? The F-35 has transitioned from a traditional strike fighter to a multi-role "sensor-shooter-node" that acts as an airborne brain for the air-ground team. Its primary value now lies in its ability to penetrate threat rings to discover targets and distribute high-quality data to the broader joint force.

What are the "three ways to win" regarding digital interoperability and communications? This approach ensures that Marines are never dependent on a single communication "pipe" by maintaining at least three pathways for message transmission. These include tactical data exchanges (Link-16), protected military SATCOM, and unclassified commercial networks like Starlink or cellular fiber.

PART III: ESSAY QUESTIONS

Instructions: Use the provided source context to develop comprehensive responses to the following prompts.

Logistics as the Limiting Factor for the Impact Force: Analyze the challenges of "contested logistics" in distributed operations. How do platforms like the CH-53K and concepts like Aviation Ground Support (AGS) attempt to mitigate these vulnerabilities?

The Cultural Shift of Mission Command: Discuss the tension between technical connectivity and decentralized execution. Why does the document argue that "centralization under stress" is a primary risk to the success of an impact force?

Aviation as Connective Tissue: How do the concepts of Distributed Aviation Operations (DAO) and Decision-Centric Aviation Operations (DCAO) redefine the Aviation Combat Element's role within the MAGTF?

Evidence-Based Adaptation: Examine the October 2025 Force Design Update. How does the decision to retain the 4th Marine Regiment and restore breaching capabilities demonstrate that the Marine Corps is a "learning organization"?

The Proxy of Modern Conflict: Using the references to the war in Ukraine found in the text, discuss why "adaptation speed" is considered a more critical metric for an impact force than raw platform performance.

Glossary of Key Terms

Term	Definition
Aviation Ground Support (AGS)	The seventh function of Marine aviation; includes the fuel, power, and infrastructure required to sustain expeditionary hubs, spokes, and nodes.
Campaign of Learning	An iterative process of war-gaming, exercises (like Steel Knight), and real-world operations used to refine Force Design concepts based on evidence.
Chaos Management	An operational philosophy assuming a permanently contested environment with overlapping threats, blurred lines between peace and war, and compressed decision cycles.
Decision-Centric Aviation Operations (DCAO)	A shift from platform-centric planning to data-centric warfighting, treating every aircraft as a sensor and data relay to achieve machine-speed decision advantage.
Digital Interoperability (DI)	The effort to connect existing and emerging sensors, shooters, and decision-makers across services and allies into a resilient, adaptive network.
Distributed Aviation Operations (DAO)	The dispersal of aviation assets across a network of hubs, spokes, and nodes to increase survivability and complicate adversary targeting.
Force Design 2030	The Marine Corps' strategic blueprint for transformation, evolving through updates to prioritize joint kill-web enablement and maritime competition.
Hub-Spoke-Node	A taxonomy for distributed operations: **Hubs** provide robust maintenance and C2; **Spokes** are austere but sustainable sites; **Nodes** are transient, time-bounded mission locations.
Impact Force	A force designed to generate disproportionate effects for the joint team by acting as the sensing, targeting, and command-and-control layer of the kill web.
Inside Force	A concept involving forces positioned within an adversary's weapons engagement zone (WEZ) to complicate targeting and deny sea control.
Kill Web	A non-linear network of sensors and shooters where any vetted platform can cue any weapon, providing resilience against the failure of individual nodes.
Project Eagle	The strategic roadmap for Marine aviation modernization, divided into three horizons: "Fight Tonight," "Bridge the Gap," and "Future Fight."
Stand-In Force (SIF)	Small, low-signature, highly mobile Marine units designed to operate and persist within an adversary's weapons engagement zone.
Weapons Engagement Zone (WEZ)	The range within which an adversary's long-range precision fires and sensors can effectively target and strike opposing forces.

SHAPING A NEW PILOT CULTURE: WYNNE AND BERKE DISCUSS THE WAY AHEAD FOR AIRPOWER

In a meeting at the 33rd Fighter Wing in early September 2013, Secretary Wynne and LtCol Berke discussed the evolving impact of 5th generation aircraft on combat operations. Wynne as Secretary of the USAF together with the Chief of Staff of the Air Force led an effort to put non-USAF pilots into an F-22 to jump start USAF thinking and to gain better joint force understanding the transition.

LtCol Berke was a key player in the effort, as a USMC pilot, he went to Nellis to train on the F-22. Lt.Col Berke is now the F-35B squadron commander for the USMC at the 33rd FW, and is the only F-22 and F-35 pilot in existence.

His background is truly unique. Suffice it to note that he has accumulated over 2800 flight hours in the F/A-18, F-16, and F-22, and F-35.

The meeting at Eglin was the first time that the formulator of the 5th generation aircraft concept had met LtCol Berke and provided them with an historic opportunity to look backwards, and more importantly forward to the evolving impact of the new aircraft on combat.

The discussion began with Wynne explaining his thinking about the necessity for the cross-assignment.

It boiled down to the fact that I believed the USAF needed to better understand and explain that 5th generation aircraft are not simply replacement aircraft for the 4th generation. I believed that bringing in pilots from other services and Air Forces might well jump start USAF thinking as well as spread the word to others.

Berke then underscored that he had come to Nellis at a good time, because the USAF was beginning to understand that the F-22 was not simply the next iteration of the Eagle and that they would have to focus more than they had on how the 5th generation would work with legacy aircraft to shape more effective combat capability overall.

Secretary Wynne had considered early on that there was an inherent advantage to leveraging legacy aircraft as the first shooters in any serious engagement to better use the stealth characteristics of the fifth generation. This means the relearning of basic pilot instinct to shoot first to protect those following. Here it is shoot from follow platforms, and save ordnance for the final fight.

Berke saw this as well.

I got to Nellis at the time when the F-22 community was beginning to really understand the necessity to better integrate the F-22 within the overall air force. When I was there, the most significant tests we were doing were integration tests.

Berke underscored that "a strike force of Raptors working with Hornets, or Eagles or Vipers are going to do better in an overall air combat effort than simply training to operate by themselves." He also highlighted that this experience was central to his work at Eglin in shaping an approach for the roll out of the F-35B to the USMC.

When asked about the evolution of the F-22 into the most lethal SEAD (Suppression of Enemy Air Defense) aircraft ever built, Berke underscored that F-16 pilots were key players in shaping thinking about this evolution for the F-22 and its contribution to the overall air combat effort.

In other words, already the cross fertilization of legacy with 5th generation aircraft have shaped a new approach to the crucial SEAD mission, one highlighted by recent Syrian events as well.

Wynne emphasized that the Berke approach was central to the "renorming of airpower" and that a key aspect of the transition is leveraging 5th generation aircraft is reshaping the sensor-shooter relationship.

The fifth generation pilots are going to have to be trained that firing first is not their core con-ops. Giving validated targets to other shooters is the 'to be' condition. This is reversing decades of training and experience where the instinct is to fire first and ask questions later.

With 5th generation aircraft you are setting up the air space for air dominance, and weapons are delivered from assets throughout the managed airspace. Without the 5th generation aircraft you have to fight your way in and expend significant effort just trying to survive. With the 5th generation aircraft you are setting up the grid to shape the offensive and defensive force to achieve the results which you seek.

Lt. Col. Berke also emphasized the core challenge of re-shaping the pilot's instincts as evident in legacy aircraft.

I am often asked to compare legacy to 5th generation aircraft and this is really difficult to do if you have not flown the aircraft. I love my F-

18 and it will always be my aircraft. But it can never be a 5th generation aircraft.

The basic way to understand the 5th generation aircraft is that it allows you to determine where in the battlespace you will fly, without the adversary setting up force barriers which need to be destroyed before I can operate a legacy fleet.

In my F-22 or F-35 I can operate in the full spectrum of combat – RF, EO, IR, etc. – and can do so with width and depth of operational reach. The fleet is core to understanding this reality.

He emphasized that the F-35 has more depth than does the F-22 in operating in a full spectrum environment.

The F-35 adds layers of depth on top of what the F-22 has because there are so many different sensors looking at any field — anything in the spectrum deep, not just the radar.

It's not just the array, it's not just the EOTS, it's not just the DAS.

It's all those things overlayed.

And so you don't just have breadth, you have huge depth in whatever part of the spectrum you want.

By flying 5th generation aircraft, Berke underscored the challenge of shifting the pilot's instincts.

As a combat pilot in legacy aircraft you are working with data to execute a mission; and you fly with wing men. In the fifth generation world, you do not have wingmen and you do not have data. You have information. The data is behind the glass and the screen provides the information.

In effect you are shifting from being a tactical asset doing tactical aircraft missions to a more strategic engagement.

And this clearly affects the direction pilot training and combat thinking must now be ingrained as a part of the 5th generation driven revolution. This must be understood in the theater command structures designing are war-winning strategy.

There's a burden now that the Raptor community is feeling, and that the F-35 community will begin to feel. The tactical aircraft is no longer just a tactical platform with strategic implications.

It is a tactical, operational, and strategic platform when it needs to be. There is an obligation now because the burden on the pilot has been lifted because the information is so high fidelity, it's so accurate, and real time, and so plentiful, that the pilot now has to see himself and view himself in a larger context than we had in the past.

In a context like Syria, the 5th generation assignment might be to retarget incoming cruise missiles to target mobile launchers as they move. 5th Generation warfare is all about movement and situation awareness.

And Lt. Col. Berke hammered home again and again that his experience with legacy and then with the F-22 and F-35 simply underscored that one was describing different historical epochs in air combat capabilities and approaches, and not simply iterative changes.

How could I possibly compare the F-35 to a F-18? I have zero criticism of the Hornet. I love that jet. The Eagle is a fantastic airplane. Those are fantastic airplanes that I know and love and will miss not flying when I retire, but it's just a disservice to both airplanes.

Such a comparison dilutes the real capability that we're getting with 5th generation and incorrectly assigns capability to an airplane that was never designed, has no capacity to do tasks that have been designed into the new generation of aircraft.

The legacy aircraft operated in a different time with a different environment, and a different world where we didn't have the expectations or climate for a tactical platform to do the things that a 5th generation aircraft is built from the ground up to do.

Wynne observed that engaging operators from other Air Forces did expose all of the Air Forces engaged in the F-35 enterprise in the dimensional change being developed right before their eyes.

As Col Berke noted, further integration of 4th gen pilots led to exponential exploitation of the 5th gen capabilities.

Wynne added:

Sadly this is not soon enough for the F-22; but it appears that the concept of an international and joint fleet of F-35's has jumped the gap in thinking. Much like repeating rifles had to overcome years of training for sharpshooting; now the task is clear: get this new generation in to the hands of operators as quickly as possible. This transformation needs the speed of the Internet and the speed of mobile that underscores the future fight.

Having witnessed the discussion and participating from time to time in the dialogue with the architect of 5th generation airpower and a key shaper of its reality, what would we conclude?

It is clear that we are at an historical turning point in the development of airpower.

If we go back in history we might note that the lesson for the air power rivalry between the U.S. and competitive air forces is rather straightforward: the technology had to be available but it also had to be successful understood and employed; not just by the operators or Pilots but by the command authorities, sometimes extending to national leadership.

The lesson on the rivalries to date is that theater and air combat leaders must adjust during the course of an air battle or war by changing strategy and tactics, be able to exploit the enemy's mistakes or weakness.

The best is to have the advantage of leveraging early the introduction of technology such as the supersonic German Jet or and early stealth designs. Aircrews must be adaptable enough to follow changing commands from leadership and also, on their own initiative, to change tactics to achieve local surprise and exploitation of a combat advantage.

A key conclusion is always to assume a reactive enemy will in time develop the necessary technology to try and mitigate any advantages. With the worldwide proliferation of weapons even a second or third world nation might have state-of-the art systems.

As the history of war in the air shows it was a constantly evolving process of human factors integrated into technology. The Cold War ended well for humanity and a lot of courageous pilots, bold leaders, and smart technologists deserve a lot of credit for this great victory.

The U.S . would be wise to remember the lessons learned and along the way the loss of very good men in the air who paid in their blood for America today to have the best technology available flown by best Air Force, Navy, and Marine aviators this country can produce.

Wynne underscored:

The challenge now is to comprehend that America and the F-35 integrated international fleet has in its arsenal the wherewithal to create conditions for peace for another generation or two. Our burden is to get on with the tasks of shaping concepts of operations to take advantage of the 5th generation aircraft and the associated new tools of combat.

Published September 18, 2013.

ABOUT THE AUTHOR

Dr. Robbin F. Laird is a strategic defense analyst whose four-decade career has been defined by a singular commitment: understanding military transformation not from Washington conference rooms but from the field, alongside the warriors who make it real.

With a Ph.D. from Columbia University, where he studied under Zbigniew Brzezinski, Laird began his career in Soviet studies during the Cold War's final decades. His academic positions at Columbia, Queens College, and Johns Hopkins provided intellectual grounding, but it was his subsequent work at the Center for Naval Analyses and Institute for Defense Analyses that shaped his distinctive analytical approach. Where others saw policy from thirty thousand feet, Laird sought ground truth from those operating at sea level.

This fieldwork ethic took him to Russia, Belarus, and Ukraine in the early post-Cold War years, conducting research that would prove prescient as those regions became central to 21st-century security challenges. But Laird's most significant intellectual evolution came as he transitioned from Soviet expertise to defense transformation analysis, recognizing that understanding how military forces adapt and innovate matters more than cataloging static capabilities.

Laird operates by General Patton's principle: "If everyone is thinking alike, someone isn't thinking." This contrarian instinct led him to the F-35 program when critics were loudest and most certain of its failure. While others catalogued cost overruns and development delays, Laird traveled globally to interview pilots, maintainers, and commanders actually working with the aircraft. His "Fifth-Generation Journey" documented not the program critics thought they saw but the transformational capability operators knew they were building. Time proved his field-based analysis correct and the critics' desk-bound certainty wrong.

Similarly, when conventional wisdom dismissed the V-22 Osprey as too expensive, too complex, too controversial, Laird embedded with the Marines and Air Force special operators flying it in combat. His tiltrotor trilogy captures what could only be learned by observing the aircraft in Iraq, Afghanistan, the Pacific, and humanitarian operations that the platform's critics fundamentally misunderstood how it was reshaping power projection and enabling entirely new operational concepts.

Based in Paris as well as the United States, Laird brings a transatlantic perspective rare among American defense analysts. He doesn't just study European security. He lives within it, understanding how alliance dynamics actually function rather than how position papers suggest they should. This positioning enables nuanced analysis of NATO burden-sharing, European defense integration, and the divergent strategic cultures shaping allied responses to Russian aggression and Chinese ambition.

His extensive field research spans European allies, Australian defense forces, and North American commands. He has participated in operational exercises, flown in military aircraft during training missions, and conducted hundreds of interviews with defense officials across continents. This global engagement grounds his analysis in comparative understanding or how different nations approach similar challenges, what the United States can learn from allies, and why coalition interoperability matters more than most Washington planners appreciate.

Laird's documentation of Marine Corps transformation from 2007 through 2025 represents perhaps his most significant contribution to defense literature. Beginning with "Three Dimensional Warriors" in 2010, he identified how the V-22 and F-35B would fundamentally reshape Marine operations before most analysts grasped their implications. Over twelve volumes and fifteen years, he has chronicled this transformation with unmatched depth and access.

What distinguishes this body of work is its foundation in sustained engagement. Laird didn't parachute in for quick visits. He returned repeatedly to MAWTS-1, embedded with 2nd Marine Air Wing units, flew in Ospreys and observed F-35 operations, interviewed everyone from commanding generals to crew chiefs. He documented the CH-53K's development from concept through Israeli operational deployment. He captured how digital native pilots think differently about networked warfare.

This sustained engagement enabled him to trace transformation's arc across administrations, budget cycles, and strategic shifts. He saw the Obama "Pivot to the Pacific," the Trump administration's embrace of great power competition, and the post-2020 acceleration of distributed operations concepts. By maintaining consistent field presence, he documented not just what changed but how and why, the decisions made, the challenges encountered, the solutions developed, the lessons learned.

While Laird's work necessarily focuses on aircraft, ships, and systems, his deeper interest lies in how organizations adapt and innovate. His MAWTS-1 volumes examine training transformation. His 2nd MAW analysis explores cultural evolution. His transformation path studies investigate how the Corps navigated from counterinsurgency optimization to peer competition preparation.

This organizational focus reflects recognition that platforms matter less than concepts, and concepts matter less than the people who develop and execute them. The F-35 is revolutionary not because of its airframe but because of how operators rethought warfare around its capabilities. The Osprey transformed power projection not through speed and range alone but through the enterprise of innovation it catalyzed. The CH-53K's significance lies not in lift capacity but in its digital architecture enabling integration with future autonomous systems.

Laird consistently emphasizes the human dimension, the leaders who championed change despite resistance, the junior Marines empowered to innovate, the instructors who trained new generations differently, the maintainers who kept revolutionary systems operational. This human-centric analysis distinguishes his work from technology-focused defense writing.

As Editor and Co-Founder of *Second Line of Defense* and *Defense.info*, and Board of Contributors member for *Breaking Defense*, Laird has built platforms amplifying practitioner voices often marginalized in policy debates. His publications prioritize interviews with operators over think tank position papers, field observations over theoretical constructs, and comparative allied analysis over America-centric frameworks.

He has gone twice a year to Australia in support of the Sir Richard Williams Foundation defence seminars and is a Research Fellow with the Foundation. He has written six books on Australian defence based on his time in Australia and his work with the Foundation.

His analytical method emphasizes what he calls "chaos management" over traditional crisis management, recognizing that contemporary security environments resist the neat categorization crisis management assumes. This framework informs his analysis of Ukraine's evolution into a system-defining global confrontation, autonomous maritime systems development, and kill web concepts in distributed operations.

His recent analysis addresses the transition from crisis to chaos management frameworks, autonomous maritime systems, and kill web evolution in distributed operations. This work builds on three decades of studying how military forces adapt to strategic surprise, technological disruption, and operational complexity.

Studying under Zbigniew Brzezinski at Columbia left lasting imprints on Laird's analytical approach. Brzezinski's integration of history, geography, and power dynamics, his recognition that ideas matter but so do geography and force, shapes Laird's refusal to separate strategic thought from operational capability. Brzezinski's skepticism toward conventional wisdom and willingness to challenge prevailing orthodoxies resonates in Laird's contrarian instincts.

Like his mentor, Laird understands that security analysis requires historical perspective, comparative frameworks, and willingness to learn from those who disagree. His Soviet studies background, though seemingly distant from contemporary defense transformation, provides essential context for analyzing authoritarian systems, understanding Russian strategic culture, and recognizing patterns in how peer competitors think about military competition.

In an era when defense analysis often privileges theoretical elegance over empirical accuracy, cost projections over capability assessment, and political positioning over strategic insight, Laird's field-based, operator-focused, comparatively-grounded approach offers essential corrective perspective.

His Marine Corps transformation chronicle provides future historians authoritative documentation of how a military service navigated fundamental strategic shift. It gives current practitioners detailed analysis of transformation pathways, challenges, and solutions. It offers allied forces considering similar transformations templates applicable to their circumstances. It reminds policymakers that successful innovation requires sustained commitment, cultural adaptation, and practitioner empowerment—not just budget authorizations.

Most importantly, Laird's work demonstrates that understanding military transformation requires getting out from behind desks, traveling to where change actually happens, listening to those making it real, and documenting their journey with intellectual honesty regardless of whether their success or failure aligns with one's preconceptions.

As he continues chronicling defense transformation into 2026 and beyond, Laird brings four decades of experience, global perspective, sustained field engagement, and unwavering commitment to letting operational reality inform strategic analysis. His voice matters because it speaks from the field, amplifies practitioners over pundits, and insists that understanding how military forces actually adapt matters more than how theories suggest they should.

Beyond credentials and publications, Laird brings to his work genuine curiosity about how people and organizations navigate change, deep respect for those serving in uniform, and conviction that getting the analysis right matters because lives and security depend on it. He writes not to build reputation or advance arguments but to document history as it unfolds and share insights that might help those making difficult decisions in uncertain times.

His twelve books on Marine Corps transformation represent not just analytical output but sustained engagement with an organization and the people transforming it, a commitment to understanding their journey, capturing their innovation, and ensuring their story gets told accurately for those who will follow.

CHRONICLING THE TRANSFORMATION PATH

Over the past fifteen years, a profound transformation has reshaped the United States Marine Corps from a force optimized for counterinsurgency operations into a digitally networked, distributed force prepared for great power competition.

Through twelve books examining different facets of this evolution, a comprehensive narrative emerges of how technological innovation, operational experimentation, and cultural adaptation have converged to redefine Marine Corps capabilities for the 21st century battlefield.

THE FOUNDATION: AVIATION AS STRATEGIC ENABLER

The transformation story begins with recognition that Marine aviation is not merely a supporting element but rather the strategic glue enabling the Corps' unique operational concepts. *Three Dimensional Warriors: The Roles of the Osprey and the F-35B* (2010) established this foundational premise, arguing that aviation provides the essential capability allowing Marines to function as "three dimensional warriors" fighting the "three block war." This work presciently identified how the V-22 Osprey and F-35B would revolutionize Marine operations, providing 360-degree situational awareness and leveraging ground forces against hybrid threats across the spectrum of conflict.

THE TILTROTOR REVOLUTION

Three volumes trace the extraordinary journey of the V-22 Osprey from controversial development program to indispensable strategic asset. *A Tiltrotor Perspective: Exploring the Experience* (July 2025) reveals the untold story of how the Osprey survived "the most vicious political and media assault in Pentagon history" to become a backbone of American military operations. Through exclusive interviews with pilots, crew chiefs, maintainers, and commanders, this account demonstrates how the aircraft that "couldn't work" became the one forces "couldn't do without."

A Tiltrotor Enterprise: From Iraq to the Future (March 2025) shifts focus from the platform itself to the broader ecosystem of innovation it enabled. The book demonstrates how the Osprey transformed the Marine Corps "from a traditional amphibious force into a rapid-deployment global strike capability," while simultaneously revolutionizing Air Force Special Operations, Navy logistics, and promising to anchor Army distributed warfare doctrine. The critical insight is that the real story isn't the machine but the enterprise, the web connecting pilots, engineers, tacticians, and industry partners who envisioned a multi-domain warfighting system turning distance into advantage.

The Role of the Osprey in the Pivot to the Pacific (July 2023) examines how the Osprey catalyzed fundamental shifts in joint maritime operations. Taking two temporal snapshots—the Obama Administration's 2013 "Pivot to the Pacific" and post-2019 developments following recognition of great power competition—the work shows how tiltrotor capabilities stimulated convergence among the Navy's distributed operations, the Air Force's agile combat employment, and Marine Expeditionary Advanced Base Operations (EABO). The Osprey emerged not just as an enabler but as a forcing function driving broader operational transformation.

FIFTH GENERATION WARFARE

The F-35 Lightning II's introduction represents perhaps the most significant capability leap in Marine aviation history. *My Fifth-Generation Journey: 2004-2018* (March 2024, updated June 2025) provides a personal chronicle of the F-35's evolution from concept through global deployment. The narrative captures perspectives from pilots, maintainers, logisticians, and government officials who navigated intense criticism and political opposition to deliver what the author characterizes as "the first true Information Age flying combat system."

The updated second edition (June 2025) extends the analysis through 2025, including Israeli operational experience. It emphasizes how the F-35 represents not an end point but the beginning of continuous evolution, a "software-upgradeable aircraft" designed for constant improvement rather than periodic major upgrades. The work explores the F-35's future role as a command node for "man-robotic wolf packs," directing autonomous sensors, weapons, and support systems, positioning it as a bridge between traditional piloted aircraft and future autonomous operations.

Critical to the F-35 narrative is recognition that the aircraft demands new thinking, not merely new equipment. Nations embracing networked, data-driven warfare approaches gain decisive advantages, while those clinging to legacy processes risk obsolescence. The F-35 program created not just a new aircraft but a new model for international defense cooperation, continuous capability development, and networked operations.

THE DIGITAL HEAVY LIFTER

The Coming of the CH-53K: A New Capability for the Distributed Force (June 2023) documents how the King Stallion represents a fundamental departure from its mechanical predecessor. The work argues the aircraft should really be designated CH-55 to emphasize how profoundly different it is from the CH-53E. Built through digital thread development and manufacturing, the CH-53K was designed with maintainability and fleet support as core program elements.

The King Stallion's capabilities create new tactical options with strategic implications. It can lift 27,000 pounds of external payload and deliver it 110 nautical miles to a high-hot zone, loiter, and return with fuel to spare. This means JLTVs and up-armored HMMWVs go to shore by air rather than slower sea lift. The platform's fly-by-wire flight controls, the first for any heavy lift helicopter, drastically reduce pilot workload and enable operations in degraded visual environments, allowing crews to focus on mission management rather than mechanical flying tasks.

Critically, the CH-53K "can operate and fight on the digital battlefield." Its digital architecture positions it to integrate with unmanned systems as the Marines expand autonomous operations, making it not just a lift platform but a node in networked distributed operations.

THE TRAINING REVOLUTION

MAWTS-1: An Incubator for Military Transformation (June 2024) co-authored with Ed Timperlake examines the Marine Aviation Weapons and Tactics Squadron One at Yuma as the crucible where new platforms become integrated capabilities. Based on visits spanning 2011 to 2024, the work traces how MAWTS-1 evolved from "Project 19" into the premier aviation training squadron in the world, keeping Marine aviation "moving forward" as threats dynamically evolve and technology fosters new operational approaches.

The book emphasizes MAWTS-1's truly multi-domain character and its role exploiting the capabilities of the V-22, F-35, CH-53K, TPS-80 radar, and Common Aviation Command and Control System (CAC2S). Through interviews with pioneers and commanding officers, the work demonstrates how "a small cadre of individuals, disappointed in their assessment of post-Vietnam Marine aviation," set a new course that fundamentally reshaped Marine aviation culture.

MAWTS-1: 2023 Visit and Interviews (December 2023) provides a focused snapshot of the command working enhanced force mobility within a joint force itself in fundamental transition. The challenge identified is conducting Marine Corps reorganization while the Navy pursues distributed maritime operations and the Air Force develops agile combat employment, all simultaneously. The work captures how MAWTS-1 trains for uncertainty, preparing the force "that might have to fight tonight" while broader transformation efforts remain works in progress.

THE STRATEGIC CONTEXT

The U.S. Marine Corps Transformation Path: Preparing for the High-End Fight (March 2022) provides comprehensive analysis of how the USMC navigated the strategic shift from land wars to peer competitor threats. Beginning with the Osprey's 2007 introduction, the work demonstrates through extensive interviews how the Marines crafted innovative solutions for great power competition's return. The book emphasizes that many coalition partners look to the USMC as a relevant benchmark, finding its size more applicable than attempting to emulate total U.S. force structure.

A critical insight is that legacy forces from land wars lack direct warfighting relevance for peer competition, requiring fundamental rethinking rather than incremental adaptation. As LtGen (Retired) George Trautman observed, "it's not the equipment that will make the Corps successful on the future battlefield — it's the Marines." This human dimension of transformation, empowering junior Marines, fostering bottom-up innovation, and developing leaders who think differently, emerges as equally important as technological advancement.

THE CUTTING EDGE

The most recent volume, *2nd Marine Air Wing: Transitioning the Fight Tonight Force* (September 2025), provides detailed examination of how Second Marine Aircraft Wing underwent remarkable evolution between 2007 and 2025, becoming a digitally driven, combat-ready force able to respond globally at a moment's notice. The work documents the shift from crisis management to chaos management frameworks, examining how platforms, sensors, and culture converged to enable distributed operations.

Central to this transformation is the emergence of "digital native" pilots, the iPad generation trained from day one in networked, integrated, data-powered warfighting. At Beaufort, South Carolina, the F-35 pilot pipeline produces aviators ready to exploit fifth-generation technology's full potential and develop tactics defining future conflict. The work demonstrates how transformation isn't only about platforms and sensors but fundamentally about culture, forward-thinking leaders, empowered junior Marines, and bottom-up innovation driving readiness.

Command and control reimagination receives particular attention, including creative fielding of mobile C2 solutions like retrofitting MV-22s as airborne nerve centers with digital communication suites, and widespread adoption of ruggedized MAGTAB tablets for real-time secure sharing across every echelon.

SYNTHESIS AND SIGNIFICANCE

Collectively, these twelve books document a transformation journey extending from 2007 through 2025, capturing how technological capability, operational innovation, and cultural adaptation combined to reshape Marine Corps aviation fundamentally. Several themes emerge across the corpus:

- Technology as Catalyst, Not Solution: New platforms, Osprey, F-35, CH-53K, created opportunities for transformation, but realizing their potential required reimagining operations, training, and culture. The hardware mattered less than how Marines thought about employing it.
- Digital Integration as Foundation: Whether the F-35's sensor fusion, the CH-53K's fly-by-wire controls, or the widespread adoption of MAGTAB tablets, digital networking emerged as the connective tissue enabling distributed operations. The shift from mechanical to digital systems fundamentally altered what became possible.
- Distributed Operations as Organizing Principle: The transformation narrative consistently moves toward

distribution—smaller, more dispersed units operating across greater distances while maintaining integration through digital networks. This represents profound departure from traditional concentration of force.

- Training and Culture as Critical Enablers: MAWTS-1's evolution and the emergence of digital native pilots demonstrate that platforms alone don't transform forces. Training institutions, cultural adaptation, and empowered junior personnel proved essential to realizing new capabilities' potential.

- International Cooperation as Strategic Asset: The F-35 global enterprise and allied interest in Marine transformation models demonstrate how capability development increasingly transcends national boundaries. International partnerships became force multipliers rather than mere political gestures.

- Continuous Evolution Over Discrete Modernization: The F-35's software-upgradeable architecture exemplifies the shift from periodic major upgrades to continuous capability development. Transformation became an ongoing process rather than a destination.

- Joint Force Interdependence: Marine transformation occurred within and was shaped by broader joint force evolution, Navy distributed maritime operations, Air Force agile combat employment, emerging multi-domain operations concepts. Service-specific transformation proved inseparable from joint operational development.

These volumes provide authoritative documentation of military transformation in action, capturing not just technological advancement but the human dimensions of organizational change, the leaders who championed innovation, the warriors who proved concepts in combat, the maintainers who kept revolutionary systems operating, and the instructors who trained new generations in fundamentally different approaches to warfare.

The collection's value extends beyond historical documentation. It offers practitioners, policymakers, and allied forces detailed analysis of transformation pathways, challenges encountered, solutions developed, and lessons learned. For coalition partners seeking to modernize their forces, these works provide templates applicable to organizations of similar scale facing comparable strategic challenges.

The journey from 2007 to 2025 demonstrates that successful military transformation requires sustained commitment across multiple dimensions, technological investment, operational experimentation, training evolution, cultural adaptation, and international cooperation. The Marine Corps experience shows transformation succeeding when organizations embrace uncertainty, empower practitioners, learn from operations, and maintain focus on preparing for the high-end fight rather than optimizing for recent conflicts.

As the 2023 end-of-course video for Weapons and Tactics Instructor course states: "It is not a question of if the Marine Corps will go into combat. It is only a matter of when." These twelve volumes document how the Corps prepared itself to answer that inevitable call with capabilities, concepts, and personnel ready for warfare's future demands.